T0269536

NEUROSCIENCE, SELFLESSNESS, AND SPIRITUAL EXPERIENCE

NEUROSCIENCE, SELFLESSNESS, AND SPIRITUAL EXPERIENCE
EXPLAINING THE SCIENCE OF TRANSCENDENCE

BRICK JOHNSTONE

DANIEL COHEN

ELSEVIER

ACADEMIC PRESS
An imprint of Elsevier

Academic Press is an imprint of Elsevier
125 London Wall, London EC2Y 5AS, United Kingdom
525 B Street, Suite 1650, San Diego, CA 92101, United States
50 Hampshire Street, 5th Floor, Cambridge, MA 02139, United States
The Boulevard, Langford Lane, Kidlington, Oxford OX5 1GB, United Kingdom

© 2019 Elsevier Inc. All rights reserved.

No part of this publication may be reproduced or transmitted in any form or by any means,
electronic or mechanical, including photocopying, recording, or any information storage and
retrieval system, without permission in writing from the publisher. Details on how to seek
permission, further information about the Publisher's permissions policies and our arrangements
with organizations such as the Copyright Clearance Center and the Copyright Licensing Agency,
can be found at our website: www.elsevier.com/permissions.

This book and the individual contributions contained in it are protected under copyright by the
Publisher (other than as may be noted herein).

Notices
Knowledge and best practice in this field are constantly changing. As new research and experience
broaden our understanding, changes in research methods, professional practices, or medical
treatment may become necessary.

Practitioners and researchers must always rely on their own experience and knowledge in
evaluating and using any information, methods, compounds, or experiments described herein. In
using such information or methods they should be mindful of their own safety and the safety of
others, including parties for whom they have a professional responsibility.

To the fullest extent of the law, neither the Publisher nor the authors, contributors, or editors,
assume any liability for any injury and/or damage to persons or property as a matter of products
liability, negligence or otherwise, or from any use or operation of any methods, products,
instructions, or ideas contained in the material herein.

Library of Congress Cataloging-in-Publication Data
A catalog record for this book is available from the Library of Congress

British Library Cataloguing-in-Publication Data
A catalogue record for this book is available from the British Library

ISBN 978-0-08-102218-4

For information on all Academic Press publications
visit our website at https://www.elsevier.com/books-and-journals

Publisher: Nikki Levy
Acquisition Editor: Joslyn Chaiprasert-Paguio
Editorial Project Manager: Megan Ashdown
Production Project Manager:
   Sujatha Thirugnana Sambandam
Cover Designer: Christian J. Bilbow

Typeset by SPi Global, India

Working together
to grow libraries in
developing countries

www.elsevier.com • www.bookaid.org

Contents

III

SELFLESSNESS AS THE KEY TO TRANSCENDENCE

IV

APPLICATIONS OF SELFLESSNESS

Acknowledgements

The authors would like to thank the University of Missouri for encouraging and supporting our research on numerous occasions. In addition, we would like to thank the numerous colleagues who have participated in a wide range of research projects that have greatly contributed to our exploration of this book's topic. Special thanks go to Dong Yoon and Shawn Christ at the University of Missouri; to Robin Hanks at Wayne State University; and to Braj Bhushan at Indian Institute of Technology (IIT), Kanpur, for their collaborations and insights. We also thank the Pew Charitable Trust and the University of Missouri's Center on Religion and the Professions who supported our initial interest in the neuropsychology of spiritual transcendence.

Dr. Johnstone is most grateful for the support he received from the Templeton Foundation and the many opportunities this provided for him to explore the relationship between the sciences and humanities. This included a delightful year with an incredibly bright (and fun) group of colleagues at the Center of Theological Inquiry studying *Religious Experience and Moral Identity* at Princeton University (with special thanks to Will Storrar and the CTI staff) and several summers with equally bright (and fun) colleagues in the *Science and Religion* program held at Oxford University (with special thanks to Stan Rosenberg and the SCIO staff).

Also, a most sincere thanks to our publisher, Elsevier, for understanding the timely importance of this topic and promoting our efforts to write a book that we hope will be of interest to a diverse academic and public audience. In particular, special thanks are extended to Megan Ashdown, our editorial project manager at Elsevier, for her continual guidance and abundant patience. We would also like to thank the reviewers of the book proposal we submitted to Elsevier, as their comments and advice have been of great assistance.

Finally, we would each like to thank our respective wives (Becky and Signe) who have been incredibly supportive throughout the process of writing this book, often reading chapter drafts and giving us invaluable comments and suggestions. We also appreciate greatly the endurance of our children (Kate and Joel; Josh and Alisha) for patiently listening, for well over 10 years, to ongoing discussions about ways to explore spiritual transcendence.

It is challenging to write a book where the authors come from very different interdisciplinary backgrounds as we endeavor to build productive bridges between the sciences and the humanities. Lastly, the authors would also like to thank one another for enduring insights and constructive compromises as we worked to bring this book to its completion. It has been an adventure!

Brick Johnstone
Daniel Cohen

THE NATURE OF TRANSCENDENCE

1

Introduction

It is almost as if there was an "off" switch for the self, buried deep in our minds, and the world's religions were a thousand different ways of pressing the switch. **(Haidt, 2012)**

1 INTRODUCTION: THE BRAIN AND SPIRITUAL EXPERIENCE

Throughout history, people have had spiritual experiences connecting them with God, divine beings, or other conceptions of a transcendent reality. These experiences were primarily viewed from the perspective of different religious faiths, but over the last several decades, the neurosciences have attempted to identify the neurophysiological correlates of spiritual experience as neuroimaging technologies have made it possible to "see" what is occurring in the brain as different activities are carried out. As a result, there has been increased attention to determine the neuroanatomical locations of spiritual experiences, looking, for example, for "spiritual centers" or a "God spot" within the brain, in attempts to discover the neural basis of diverse religious beliefs. However, the specific relationships that exist among cerebral regions, neurophysiological activities, and specific spiritual experiences remain unclear.

One problem that has been pointed out is that our ability to understand that nature of religious and spiritual experiences has remained relatively limited because the majority of neuroscientific studies have focused primarily on identifying the neuroanatomical correlates of such experiences (Schjoedt, 2009; Seitz & Angel, 2012). Most studies have attempted to identify precisely "where" in the brain spiritual experiences occur, but all human thoughts, sensations, emotions, behaviors, and experiences are brain based. Knowing that "spiritual" experiences are connected to specific brain regions does not address their experiential qualities. As a result, there is increasing interest in identifying "what" interrelated processes are occurring in the brain during spiritual experiences.

However, just as neuroscientific study has been limited by focusing on the neuroanatomy of spirituality, it has been equally limited by the lack of any universal agreement about what constitutes "spirituality." Scientists and scholars from a variety of disciplines (including the social sciences and the humanities) have developed very divergent definitions and measures of various religious and spiritual experiences that are based on their different theoretical backgrounds. For example, over the past 30 years, numerous researchers have investigated a variety of spiritual experiences using various terms including religion, religiosity, religiousness, intrinsic religiosity, extrinsic religiosity, spirituality, spiritual transcendence, transcendence, mysticism, mystical experience, unitive experience, spiritual meaning, values, religious belief, daily spiritual experience, forgiveness, religious coping, spiritual coping, and religious/spiritual coping. As such, it remains difficult to identify the underlying neurophysiological correlates of spiritual experiences, as there is no consensus on what constitutes a "spiritual experience." One problem is the difficulty of using scientific methods to study what have been traditionally considered religious, philosophical, and theological conceptions. What is needed is a way to identify common concepts that have meaning to the neurosciences, the social sciences, and the humanities and that can be studied using both objective and subjective methodologies. However, identifying concepts that have useful theoretical importance for all these scholarly domains has proved difficult.

2 THE PARIETAL LOBES, SELFLESSNESS, AND SPIRITUAL EXPERIENCES

Although spiritual and religious experiences have generally been shown to be processed throughout the entire brain, empirical studies are

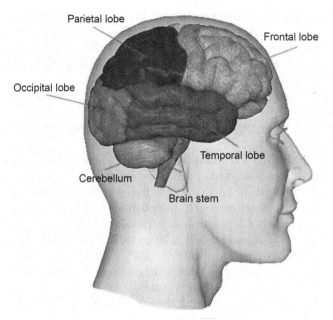

FIG. 1   The right hemisphere of the human brain and the four lobes of the neocortex, including the parietal lobe.

often suggesting that some specific spiritual experiences are associated with reduced activity in the right hemisphere of the brain generally and more specifically within the right parietal lobe (see Fig. 1). Of potentially greater importance, these studies are also suggesting that these spiritual experiences are associated with a decreased sense of self.

For example, neuroimaging studies of both long-term practitioners of Buddhist meditation and Franciscan nuns engaging in centering prayer were carried out by Newberg, d'Aquili, and colleagues 20 years ago. Their neuroimaging results indicated that both of these religiously based ritual activities, meditation for the Buddhists and centering prayer for the nuns, were associated with decreased parietal lobe activity (Newberg, Alavi, Baime, Mozley, & d'Aquili, 1997; Newberg, d'Aquili, & Rause, 2001; Newberg, Pourdehnad, Alavi, & d'Aquili, 2003). Importantly, individuals from each group also reported having experiences that were neuropsychologically very similar. That is, both the meditators and the nuns reported that at the peak of their respective ritual practices, they each had a potent spiritual experience that was accompanied by a loss of their individual sense of self. However, this phenomenologically similar loss of the sense of self was interpreted differently by the practitioners of these two different religious traditions. The Buddhist meditators described their experience of eradicating the false sense of self as achieving a momentary glimpse of the deepest truth, while the nuns said they had attained spiritual communion with God in conjunction with their loss of self. Importantly, individuals in both groups showed similar neurophysiological patterns that accompanied their strong sense of the loss of the individual self, although these experiences were understood differently based on the participants' respective religious traditions and backgrounds. In other words, their respective rituals both generated a similar neuropsychological experience involving the experience of a loss of the sense of self, despite their varying interpretations (i.e., Buddhist versus Christian) of what had occurred.

Other studies have indicated that the right hemisphere and parietal lobe specifically are related to spiritual experiences drawing on evidence from individuals with brain disorders. By correlating neuropsychological tests with self-report measures of spirituality, it has been found that injury to the right hemisphere and parietal lobe leads to significantly higher levels of spirituality. Such relationships have been identified in several studies involving individuals with various types of brain injury or disease, including traumatic brain injury (Johnstone & Glass, 2008; Johnstone, Bodling, Cohen, Christ, & Wegrzyn, 2012; Johnstone, Bhushan, Hanks, Yoon, & Cohen, 2016), epilepsy (Johnstone et al., 2014), Parkinson's disease (Butler, McNamara, & Durso, 2010, 2011), and brain tumors (Urgesi, Aglioti, Skrap, & Fabbro, 2010). The results obtained have consistently indicated that there is a relationship between decreased parietal function and heightened spiritual experience, with several of these studies concluding that spiritual transcendence is associated with reduced self-orientation that is processed in the right hemisphere.

Using a different approach to study relationships between the brain and spiritual experiences, other studies have used electromagnetic brain stimulation to elicit spiritual experiences in healthy individuals. Specifically, a technology called transcranial magnetic stimulation (TMS) can be used to either inhibit or enhance neurophysiological activity in specific, targeted regions of the brain. One study found that when the neural activity in the inferior parietal lobes was *inhibited*, participants reported having spiritual experiences (Crescentini, Aglioti, Fabbro, & Urgesi, 2014). In another study, when right inferior parietal lobe activity was *enhanced* using TMS, the spiritual orientation of participants was reduced (Crescentini, Di Bucchianico, Fabbro, & Urgesi, 2015), reaffirming the inverse relationship between reduced right parietal activity and the occurrence of spiritual experience.

These studies and others raise the question of how *inhibition* of right parietal lobe functioning may be related to spiritual experiences. If anything, one might initially suspect that increased neurophysiological activity would be necessary to create the awe and wonder reported as common elements of spiritual experiences. Yet, neuroimaging studies have consistently shown that spiritual experience is associated with a relative reduction of neurophysiological activity in the right hemisphere of the brain and particularly the parietal cortex (Barnby, Bailey, Chambers, & Fitzgerald, 2015). Specifically, why does the reduced neurological functioning of the right parietal lobe lead to increased spiritual experiences? This question, in part, fueled our interest in writing this book. Jonathan Haidt, in the quote cited at the beginning of the chapter, provides an important hint. It suggests, in our words, that certain neuropsychological processes are turned off through religious practices in order to facilitate the loss of the sense of self, which thereby enhances spiritual connection with the divine, nature, or the cosmos.

3 SELFLESSNESS: BRIDGING NEUROSCIENCE AND THE HUMANITIES

We argue that the neuroscientific study of spirituality has been limited due, in part, to the inability to identify a common construct that has relevance to both the sciences and the humanities. To advance these fields, it is necessary to identify common concepts that have meaning to the natural sciences, the social sciences, and the humanities. However, identifying concepts that have useful theoretical importance for all these scholarly domains has proved difficult.

We propose that the study of the "self" and "selflessness" may serve as bridges that facilitate a better collaboration between the sciences and humanities in the study of spiritual transcendence. Consistent with the thoughts of neuroscientist Patrick McNamara, author of *The Neuroscience of Religious Experience* (2009), we concur that,

> The Self is one key to all of the fundamental functional aspects of religion and it therefore behooves us to examine religious practices that are specifically aimed at the Self. I think it is fair to say that all religions target the Self for transformation, or better religion seeks to channel transformations of the Self into a prosocial direction by constructing a unified executive form of the Self. *(McNamara, 2009, p. 147–148)*

The philosopher Daniel Dennett would likely not capitalize "Self" as McNamara does in connecting religion with self-transformation, as he conceptualizes the "self as a center of narrative gravity" (Dennett, 1992). Dennett explores the phenomenological experience of self in great detail, and he sees the self as a useful tool for connecting psychology and philosophy by understanding the nature of experiences of loss of the sense of self (see discussion in Simpson, 2014) with which we agree. We believe that our research on the loss of the sense of self and its relationship with spiritual transcendence can extend insights in both neuropsychology and religious studies and in particular facilitate stronger conversations between these areas and also with several other disciplines and areas of research. The renowned psychologist and philosopher William James had already begun bridging disciplinary orientations by the beginning of the 20th century in several useful ways. He categorized different empirical senses of the self and helped initiate the formal study of the psychology of religion, in large part through his own exploration of intense mystical states. James connected the human potential of losing the sense of self with experiences of profound spiritual transcendence, understanding that these experiences occur in various religious contexts. As James said,

> The overcoming of all usual barriers between the individual and the Absolute is the great mystic achievement. In mystical states we both become one with the Absolute and we become aware of our oneness. This is the everlasting and triumphant mystical tradition, hardly altered by difference of clime or creed. *(James, 1902, p. 362–363)*

In neuroscience in general, there has been increased focus on the neuropsychological processing of the self over the past 25 years. Research has increasingly indicated that the "self" is best conceptualized as being constructed from integrated neuropsychological processes that create a "sense of self," rather than traditional philosophical notions of the self as a permanent entity. These insights have been developed through fascinating case studies of individuals with focal neurological impairments in the right hemisphere that have identified a wide range of conditions known as "disorders of the self." These include asomatognosia (i.e., denying that part of the body is one's own), mirror misidentification disorder (i.e., the inability to recognize oneself in the mirror), and anosognosia (i.e., the inability to recognize the nature of one's strengths and weaknesses). In addition to case studies, recent empirical research is further indicating that brain networks housed primarily in the right hemisphere are responsible for processing information that is specific to the self (i.e., self-oriented), including physical, psychological, and autobiographical selves.

We further note that while the "self" has only been studied in a systematic fashion in the neurosciences for the past several decades, in the humanities, the "self" has been a focal point of interest for thousands of years. For example, allusions to the loss of the sense of self associated with spiritual experiences are found in the literature of a variety of religious traditions from a wide assortment of geographic regions, historical time periods, and diverse cultural backgrounds. Specifically, religious faiths from across the globe with widely divergent belief systems have described selflessness (i.e., loss of the sense of self) as a core component of spiritually transcendent experiences—experiences that have been sought after by spiritual seekers over the past 3000 years or longer. Thus, it appears that the sciences and the humanities share a common interest in studying the "self" and its loss, although from different orientations.

We find, in agreement with many others, that the "self" is better understood as a process rather than a fixed thing, aligning us with the "bundle theorists" of philosophy, religious studies, and neuroscience, discussed briefly by Anil Ananthaswamy (2015) in his book *The Man Who Wasn't There*. When the self is conceived of as a process, it is easier to comprehend how spiritual experiences may be related to a "decreased sense of self" and experiences of profound spiritual transcendence with "selflessness." We propose that by studying spiritual transcendence through both neurological and religious perspectives of the "self," scholars from both the sciences and humanities can work collectively to improve our understanding of the nature of spiritual transcendence.

4 THE MODEL: MAPPING THE SELF, SELFLESSNESS, AND SPIRITUAL TRANSCENDENCE

We have developed a model of the relationship between spiritual transcendence and selflessness using evidence from neurological and neuropsychological studies of the right hemisphere/right parietal lobe of the brain, genetic studies of twins, recent work in neuroendocrinology, cross-cultural studies on spirituality and personality, and the examination of religious and nonreligious descriptions of profound experiences of transcendence. We are proposing that self-transcendence is a universal neuropsychological process for all individuals that can be or has been experienced by all individuals across the globe and throughout time. However, the manner by which experiences of self-transcendence are interpreted is influenced by context and one's background. In the context of religious understanding, Christians may experience a transcendent connection with God, Buddhists the truth of emptiness, and atheists a profound connection with nature or the cosmos. In order to understand how culture and religion influences the interpretation of such experiences, we rely on the wisdom of religious texts, sayings, and beliefs. In the following chapters, we present information from both neuroscience and the humanities and propose a universal neuropsychological model of spiritual transcendence that has relevance to both these disciplinary domains. Following are the basic elements of our model:

- The right hemisphere in general and right parietal lobe specifically process information to create a "sense of self."
- This sense of self can be inhibited through injury or disease to those parts of the brain related to processing a sense of self (i.e., right hemisphere/right parietal lobe), leading to a decreased sense of self.

- This decreased sense of self can also be brought about through behavioral and ritual practices such as prayer, meditation, and asceticism.
- A decreased sense of self (i.e., increased selflessness) can produce an associated unitive experience of a wider connection beyond the self.
- This self-transcendence can be experienced as spiritual transcendence and interpreted differently depending on personal, cultural, and religious contexts (e.g., *unio mystica*, *kensho*, and *tawhid*).

To better understand spiritual transcendence, it is important to explore both the neural foundations of such occurrences and human responses, interpretations, and understandings arising from such experiences. To do this, we need to combine the precise methodologies of science with the deep insights of the humanities. Neuroanatomy and neuropsychology help provide scientific explanations for how a reduced sense of self or selflessness may be produced within the brain during a spiritually transcendent experience, whereas religious descriptions similarly stress the importance of reducing or minimizing focus on the self in order to transcend oneself to connect with higher powers, cultivate wisdom, and achieve a deeper awareness of the meaning of life.

Needless to say, the human brain is immensely complex, and there are many parts of the brain that are involved that interact in complicated ways making it extremely difficult to understand the sense of "self." We are not aiming to explore every possible area of the brain that could relate to understanding or constructing all facets of the self, but rather focus on one particular hemisphere of the brain (i.e., the right hemisphere and especially the parietal lobe), strongly associated with both spiritual experiences and a basic loss of the sense of self. We acknowledge that the rush to comprehend all the dynamic interactions of different parts of the brain vis-a-vis the "self" has often made the development of more basic understandings of the relationship between spiritual experience and brain function overly complicated, at times obscuring understanding of this association rather than clarifying it. This book develops a model of selflessness based on the basic loss of the sense of self (i.e., reduction or curtailment of the awareness of self), exploring the connection between selflessness and the propensity for transcendent spiritual experiences.

5 THE EVOLUTION OF THE PARIETAL LOBES, THE SELF, AND SELFLESSNESS

Although our neuropsychological model of spiritual transcendence focuses heavily on the perspectives of the neurosciences and humanities, recent theories from other natural sciences also have relevance to our model. For example, it is readily apparent that brain size has increased in humans over the course of evolution and that the human brain is comparatively larger than what is found in other primates. These increases can be measured and compared using the encephalization quotient (EQ), a form of comparative brain measurement that is calculated by dividing brain size by body mass. If restricted to a survey of monkeys, apes, and humans and scaled to human body size, the EQ shows that human brains are about three times larger than what would be predicted comparatively (Klein, 2017). During human evolution, encephalization has occurred in spurts with a sharp increase seen around 2 million years ago, around the time of the proliferation of the *Homo* genus and its movement out of Africa. Another significant boost in size appears to have occurred around 600,000 years ago, and an even stronger gain occurs between 130,000–140,000 years ago in *Homo neanderthalensis*, with the largest relative increase seen in the development of modern *Homo sapiens* (see Klein, 2017 for discussion and additional references; also see Fig. 2).

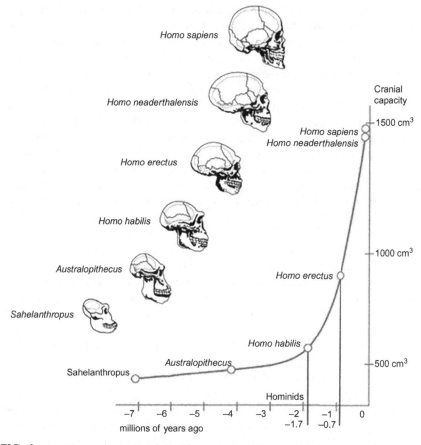

FIG. 2   Increases in brain size during hominin evolution (includes modern humans, extinct human species, and other ancestors).

Although the overall size of the human brain has increased over time, most attention has been focused on the relative growth of our frontal lobes. This is likely because the frontal lobes have been associated with the development of executive skills, including the capacity to reason, plan, and organize (i.e., those abilities that separate humans from other species). However, our parietal lobes have also significantly increased relative to other species, although more recently than other regions of the brain. This increase in the size of the parietal lobes raises questions about the development of associated neuropsychological abilities and how they may relate to unique human experiences.

More specifically, the most significant development of the parietal lobe appears to have occurred very recently as brain size abruptly increased with the development of modern human species (i.e., *Homo sapiens* and their contemporaries *Homo neanderthalensis*). At this time, it also appears that there was a significant reorganization of the neocortex that facilitated the development of much greater parietal surface area. However, this specific change is particularly pronounced in *Homo sapiens,* but is not seen in *Homo neanderthalensis*, and it is characterized by a "bulging of the parietal region" (Bruner, Preuss, Chen, & Rilling, 2017, p. 1053).

In modern humans, the expansion of the parietal surface occurs in the first year of life, during a "morphogenetic stage" not seen in Neanderthals or in chimpanzees, humans' closest living relatives (Bruner et al., 2017). Thus, a major difference between human and chimpanzee brains is the pronounced longitudinal expansion of the parietal lobe. Leading experts in the study of paleoneurology have noted these differences and have called them "a key feature of modern human evolution" that comprises the "main difference" between human and chimpanzee brains (Bruner et al., 2017, p. 1058). In addition, the intraparietal sulcus (i.e., furrow) is much larger and far more complex in humans in comparison with nonhuman primates (Bruner, Amano, Pereira-Pedro, & Ogihara, 2018).

In terms of our model of selflessness and its implications for the parietal lobe of the brain, evidence from the human fossil record shows that the enhanced development of the human parietal lobe is a relatively recent development in human evolution. This late occurrence is likely significant, as the right parietal lobe of the brain has been strongly implicated in the sense of self, and when its neural input is reduced, the loss of the sense of the self, or increased selflessness, heightens the potential for spiritual experience.

We believe the recent enlargement of the parietal region in modern human evolution has significance for the model presented in this book. Human fossils dating back to 150,000–200,000 years ago, including Neanderthals, do not show this bulge in the parietal region (Bruner & Pearson, 2013; Bruner et al., 2017). The expansion of the human parietal lobe is a very recent occurrence in human evolution, and it appears to coincide with the abrupt elaboration and increasing sophistication seen in archaeological assemblages in the production of tools, artifacts, and art that occurs in the middle to late Stone Age (see, e.g., Klein, 2000). The parietal cortical area is involved in many complex cognitive functions; is particularly important for the visuospatial integration of the eye-hand system and body integration; and is also involved in self-awareness, egocentric memory, and mental imagery (Bruner & Iriki, 2016; Bruner et al., 2018). The recent evolutionary change in the parietal lobe of modern humans also occurs shortly before the appearance of early human burials, a behavior that may have happened among some Neanderthal groups, and is seen as increasingly associated with modern *Homo sapiens (sapiens)* fossils, strongly suggesting the development of enhanced expressions of spiritual belief with regard to life and death. While speculative, it is not difficult to imagine that experiences of selflessness were also becoming enhanced as well and perhaps began to be sought after ritually, something that is seen in many extant hunter gatherer and shamanistic cultures.

6 OUTLINE OF THE BOOK

We argue that it is only by considering information from the neurosciences, natural sciences, social sciences, and the humanities that we can adequately develop a neuropsychological model of spiritual transcendence. It is only by integrating such diverse information that we can better understand and ultimately impact the human experience of spiritual transcendence.

We will explore our model of the self, selflessness, and spiritual transcendence through the following sections and chapters. In the first section of the book, "The Nature of Transcendence," we introduce our model of the loss of the sense of self and attendant selflessness based on right hemisphere and right parietal lobe processing.

Specifically, in this chapter, we are pushing for more collaboration between the sciences and the humanities in the study of spiritual transcendence, suggesting that enhancing their dialogue and partnership will be a benefit to each, and provide a rationale for focusing on "spiritual transcendence" as a unique construct. Here, we introduce the reader briefly to our model about the relationship between the self, selflessness, and transcendence and lay out our intention to focus on the right hemisphere/right parietal lobe of the brain. To further explain this orientation to the relationship between the sense of self, selflessness, and spirituality, we briefly discuss recent findings on the evolution of the human brain that support our model and its focus.

In Chapter 2, we present additional information showing that spiritual transcendence is a specific, empirically verifiable construct that is worthy of scientific investigation and related in particular to a greatly reduced sense of self. Here, we discuss how both the scientific and the religious study of experiences of spiritual transcendence have been limited by weaknesses in defining and measuring these experiences. We provide personal and religious descriptions of experiences of spiritual transcendence that provide clues to their nature. We note that spiritual transcendence has been empirically studied since the turn of the 20th century, starting with William James and continuing with psychological research that identifies the loss of self as a core phenomenological characteristic of spiritual transcendence. Psychological research is reviewed that indicates that spiritual transcendence may be best conceptualized as a personality trait, that it is psychometrically valid, and that it is a unique and stable construct. We also present research that indicates that spirituality and "spiritual transcendence" appear to be genetically inherited and may be more stable over time than other common personality traits.

In Section 2 of the book, "The Self and Selflessness," we present neuroscientific research that indicates that both the "self" and "spiritual transcendence" are related to specific neurological regions and neuropsychological processes, with an emphasis on demonstrating that the self is more of a neuropsychological process than an entity.

Chapter 3 reviews famous case studies that show how specific, localized brain lesions can help us understand how specific brain networks are associated with specific neuropsychological abilities (e.g., frontal lobes and personality, left frontal lobe and expressive language, and hippocampus and memory). Additional anecdotal case studies are presented that show how the right hemisphere in general and right parietal lobe specifically are associated with "disorders of the self." This discussion includes examples of individuals with neurological injuries who deny that the left side of their body is their own, who cannot identify themselves in the mirror (although they can identify others), and who are completely unaware that they have suffered catastrophic injuries. Other psychiatric disorders (i.e., delusional misidentification disorders) are presented that indicate that individuals with primarily right hemisphere dysfunction exhibit impairments in relationship aspects of the self (i.e., how the self is related to others). These include cases in which individuals will claim that their family members are actually impostors or that they have existing relationships with complete strangers. This information demonstrates how scientists initially came to understand how the brain processes a "sense of self" and how it can be lost.

Chapter 4 presents empirical research that indicates that the right hemisphere creates a sense of a physical self (e.g., face, body, and voice) and a psychological self (e.g., personality characteristics). This includes studies that show how the sense of self can be altered so that individuals can believe that a rubber hand is part of the self, that an amputated arm has been reattached, or that the mental sense of self can be experienced as being in a different space than the physical sense of self (i.e., out-of-body experiences). Information is also presented that shows how advanced technologies (e.g., TMS) can inhibit the functioning of the right hemisphere to create a decreased sense of the self. The information presented indicates that rather than conceptualizing the neurological "self" as a stable entity, it is best to consider it as a set of integrated neuropsychological processes that create a "sense of self." Chapters 3 and 4 set the stage for the argument that a decreased sense of self can serve as a neuropsychological foundation for spiritual transcendence.

Chapter 5 presents empirical studies that show that decreased right hemisphere/right parietal lobe functioning and by implication increased selflessness are associated with increased spiritual transcendence for individuals from multiple religions and cultural backgrounds. These studies show that individuals with right-hemisphere disorders report increased spiritual transcendence and that individuals engaged in religious/spiritual practices who demonstrate decreased neurophysiological functions in the right hemisphere experience a decreased sense of self (or increased selflessness), all of which reinforces our model of a neuropsychological foundation of spiritual transcendence.

In the third section of the book, "Selflessness as the Key to Spiritual Transcendence," we present our neuropsychological model of spiritual transcendence and evidence from the humanities to support our model.

Chapter 6 shows how "selflessness" has been expressed as a core element of spiritual transcendence across diverse religions for thousands of years. We explore the relationship between profound experiences of transcendence that are accompanied by the loss of the sense of self expressed in a variety of religious and nonreligious texts and tenets. We argue that such experiences of selflessness share certain common underlying neuropsychological components as the perceived self becomes eradicated or obscured, whether described as a temporary disappearance or by merging with higher powers or with the entire cosmos. Here, we also point out that experiences of selfless transcendence can also occur in those who do not profess any religious affiliation, including those who identify as agnostic or atheist.

Chapter 7 expands our neuropsychological model of spiritual transcendence explored from neuropsychological, emotional, and cultural perspectives. This chapter examines recent research investigating spiritual transcendence and neurochemicals (e.g., oxytocin and dopamine), spirituality and selflessness, and how culture can influence neural organization and spiritual experience.

Finally, Section 4, "Applications of Selflessness," the last section of the book, contains a concluding chapter that offers some perspectives on how scientific attitudes toward spiritual transcendence have changed.

Here, we make suggestions for future collaborations between the neurosciences, social sciences, and humanities.

Chapter 8 addresses ways in which the neurosciences and the humanities can work together more productively by recognizing the effectiveness of their engagement with one another, and in this context, we discuss future applications of our model for research, religious practices, and health care.

7 CAVEATS ON THE MODEL PRESENTED

Before you read further, it is necessary that we provide a few caveats.

We acknowledge that spiritual transcendence is indeed a complicated experience that is related to various neurological networks, neurotransmitters, neuropsychological processes, individual differences, cultural contexts, and religious influences. We believe that the study of spiritual transcendence initially requires the simplification of terms and models in order to speak to a broad range of researchers and clinicians from both the humanities and the neurosciences. The vast majority of academic disciplines tend to focus in detail from their own perspectives, often making it very difficult for different disciplines to converse productively and collaborate with one other. For this reason, we have omitted some technical details and theoretically complex conceptions in our analysis in order to engage with a wider audience interested in exploring the nature of spiritual transcendence. By working to combine scientific, religious, spiritual, and philosophical perspectives without alienating different viewpoints, we hope to expand interest in this area of inquiry and encourage better multidisciplinary collaboration in order to involve a wider range of interested parties in the conversation, as spiritual transcendence represents a complex human experience that is difficult to study and comprehend.

Throughout the book, we consistently indicate that self-orientation processes are *primarily* related to the right hemisphere in general and the right parietal lobe specifically. We acknowledge that different aspects of the self (e.g., proto-, core-, physical, psychological, autobiographical, and narrative) are processed throughout the entire brain, including the right frontal lobe and temporal lobes, as well as the left hemisphere and subcortical structures. With this in mind, we again reiterate that we are more interested in identifying the neuropsychological processes associated with spiritual transcendence, rather than where they originate in the brain.

Just as we acknowledge omitting certain technical information in our discussion of neuroscientific concepts, we recognize that we cannot cover every variation of spiritual transcendence found in every religious text or faith tradition from the perspective of the humanities. We acknowledge that we are attempting the near impossible: describing, defining, and measuring spiritual transcendence. Thus, we speak in general terms when describing the tenets, beliefs, and histories of various religions and faith traditions. While there is little uniformity in specific beliefs within religious traditions, let alone across religions, the experience of spiritual transcendence associated with a loss of the sense of self is something that occurs throughout.

We note that it is quite challenging to attempt to compare the spiritual transcendence of individuals from (so-called) monotheistic religions (e.g., Judaism, Christianity, and Islam), polytheistic religions (e.g., Hinduism and indigenous and tribal groups), atheistic traditions (e.g., Buddhism), and nonbelievers (e.g., human secularists, atheists, agnostics, and skeptics). We also acknowledge that individuals who do not ascribe to any religious beliefs or convictions also report having transcendent experiences, which are commonly described in terms of a selfless connection with nature, the universe, etc. In fact, we believe this raises one of the more interesting questions of our time: are the spiritually transcendent experiences of nonbelievers similar in any way to those of believers (from many types of traditions) in terms of neuropsychological experience? We believe they are.

Our book builds upon the efforts of many scholars from the sciences and humanities who have preceded us and to whom we are thankful. Just as we have benefited from their efforts, it is our hope that future researchers of many backgrounds will build upon the ideas that we present here. With these thoughts in mind, we encourage you to consider the following presentation with an open mind, which will hopefully lead to more fruitful collaborations and better understanding of the nature of spiritual transcendence.

2

The Nature of Spiritual Transcendence

What I now realize, from my study of the different religious traditions, is that a disciplined attempt to go beyond the ego brings about a state of ecstasy. Indeed, it is in itself ekstasis. Theologians in all the great faiths have devised all kinds of myths to show that this type of kenosis, or self-emptying, is found in the life of God itself. They do not do this because it sounds edifying, but because this is the way that human nature seems to work. We are most creative and sense

© 2019 Elsevier Inc. All rights reserved.

other possibilities that transcend our ordinary experience when we leave ourselves behind. There may even be a biological reason for this. (Armstrong, 2004, p. 279)

1 DIFFICULTIES IN DESCRIBING EXPERIENCES OF SPIRITUAL TRANSCENDENCE

Over the years, scholars across diverse academic disciplines have attempted to better understand and explain experiences of spiritual transcendence. This has proved difficult as there are inherent problems both in defining and measuring such occurrences. Although experiences of spiritual transcendence are often regarded as ineffable, inexpressible, and overwhelming, many individuals have nonetheless grappled with ways to express and describe powerful experiences of transcendence. As a result, numerous phrases have been used to designate these experiences, including religious experience, spiritual experience, transcendent experience, mystical experience, and unitive experience. However, universally agreed-upon definitions of spiritual and religious experiences have not been developed, which has made it difficult to study the neuroanatomical or neuropsychological foundations for these experiences in any methodological way.

How should we measure ineffable spiritual experiences that are by definition impossible to express? Going back to the middle ages, Meister Eckhart (c.1260–1328), the German theologian and Christian mystic, described how the potent transcendent experiences of the biblical prophets could not be put into words:

> The prophets walking in the light ... sometimes were moved to return to the world and speak of things they knew ... thinking to teach us to know God. Whereupon they would fall dumb, becoming tongue tied ... they were dumb because the hidden truth they saw in God, the mystery they found there, was ineffable. *(Meister Eckhart; quoted in Stace, 1960a, p. 287)*

It is possible that the inexpressible or ineffable qualities of spiritually transcendent experiences may be related to how our brains are wired. This may explain in part why experiences of heightened spiritual awareness often involve a loss of the sense of self and typically produce strong emotional qualities that are difficult to put into words.

Lord Alfred Tennyson (1809–92), who was the Poet Laureate of Great Britain and Ireland, recounted similar experiences. Although raised as a Christian and expressing agnostic views toward the end of his life,

Tennyson reveals that throughout his lifetime, he had formative transcendent experiences that he found difficult to describe, despite his later status as a gifted writer and poet. In a letter, Tennyson described his lifelong transcendent experiences as those times where he lost his sense of individual self and how these experiences remained "utterly beyond words":

> A kind of waking trance—this for lack of a better word—I have frequently had, quite up from boyhood, when I have been quite alone … All at once, as it were out of the intensity of the consciousness of individuality, individuality itself seemed to dissolve and fade away into boundless being, and this was not a confused state but the clearest, the surest of the sure, utterly beyond words—where death was an almost laughable impossibility—the loss of the personality (if so it were) seeming no extinction but the only true life. *(quoted in James, Varieties of Religious Experience; also cited in Stace, 1960a, p. 119).*

Similar descriptions of spiritually transcendent experiences are described by individuals from many different religious backgrounds. For example, Martin Buber (1878–1965), an Austrian-born Jewish philosopher best known for his book *I and Thou*, deeply explored the relationship between humanity and the divine. Buber had mystical experiences throughout his life, and he is a good example of how a mystic comes to understand experiences of spiritual transcendence through the lens of the doctrines of their faith tradition. Although he had a profound loss of his sense of self during these spiritually intense occurrences, over time, Buber would come to reject the possibility that he was experiencing true union with God, as he realized that this interpretation defied Jewish theological understandings. Despite this theological reorientation, which Buber acknowledges in the quote below, he still describes his mystical experiences as connected to a complete loss of self and a profound experience of "undifferentiated unity":

> Now from my own unforgettable experience I know well that there is a state in which the bonds of the personal nature of life seem to have fallen away from us and we experience an undivided unity. But I do not know—what the soul willingly imagines and indeed is bound to imagine (mine too once did!)—that in this I had attained to a union with the primal being or the godhead. That is an exaggeration no longer permitted to the responsible understanding. Responsibly—that is, as a man holding his ground before reality—I can elicit from those experiences only that in them I reached an undifferentiated unity of myself without form or content. I may call this an original prebiographical unity and suppose that it is hidden unchanged beneath all biographical change, all development and complication of the soul. Nevertheless, in the honest and sober account of the responsible understanding this unity is nothing but the unity of this soul of mine, whose 'ground' I have reached, so much so, beneath all formation and contents, that my spirit has no choice but to understand it as groundless. *(Buber, 1947, Eng. Trans., pp. 24–25; quoted in Zaehner, 2016, p. 18)*

I. THE NATURE OF TRANSCENDENCE

Jane Goodall, one of the world's foremost primatologists, discussed a moment of profound spiritual transcendence she had in the forests of Africa and her difficulty in putting this experience into words:

> Lost in awe at the beauty around me, I must have slipped into a state of heightened awareness. It is hard—impossible really—to put into words the moment of truth that suddenly came upon me then. Even the mystics are unable to describe their brief flashes of spiritual ecstasy. It seemed to me, as I struggled afterward to recall the experience, that *self* was utterly absent: I and the chimpanzees, the earth and trees and air, seemed to merge, to become one with the spirit power of life itself. *(Goodall & Berman, 1999, p. 173)*

If names of individuals were not indicated in the quotes in this section, it might be difficult to determine which of the quotations above was from a renowned scientist, an agnostic poet laureate, or a Jewish philosopher. What we see, in these few examples, is the difficulty in coherently expressing the nature of spiritually transcendent experiences and the attendant loss of the sense of self that accompanies them. Given the ineffable qualities of spiritual transcendence, the scientific study of profound spiritual experiences remained limited until attempts were made to identify common, measurable characteristics of "religious experience" in a systematic way at the end of the 19th century.

2 WILLIAM JAMES AND THE SCIENTIFIC STUDY OF SPIRITUAL TRANSCENDENCE

The first psychologist to study experiences of spiritual transcendence in a methodological manner was William James. He was the first professor of psychology in this country (at Harvard University) and was largely responsible for developing the field of the psychology of religion in the early 20th century. In his seminal work, *The Varieties of Religious Experiences* (1902), James expressed his view that the spiritually transcendent experiences of peoples from all religious backgrounds were similar and based on common psychological processes. He wrote the following:

> In mystic states, we both become one with the Absolute and we become aware of our oneness. This is the everlasting and triumphant mystical tradition, hardly altered by differences of clime or creed. In Hinduism, in Neoplatonism, in Sufism, in Christian mysticism, in Whitmanism, we find the same recurring note, so that there is about mystical utterances an eternal unanimity that ought to make a critic stop and think and that bring it about that the mystical classics have, as been said, neither birthday nor native land. *(William James, Varieties, quoted in Harmless, 2007, p. 31)*

James was more interested in studying the powerful spiritual experiences of individuals than in specific theological beliefs or general religious practices (e.g., prayer and rituals) of different faiths. He maintained that profound spiritual and mystical experiences must be regarded as genuine and true experiences and argued that these occurrences showed strong evidence of cross-cultural and historical similarity and often constituted the spiritual basis of different religious traditions. In a letter he wrote while in Edinburgh delivering his famous Gifford Lectures (later published as *The Varieties of Religious Experience*), James referred to "the mystical experience of the individual" as "the mother sea and fountainhead of all religions" and said that "all theologies and all ecclesiasticisms are secondary growths superimposed" (cited in Forsyth, 2003, p. 110).

As one of the first and most prominent scientists to study the psychological foundations of religious experiences, James understood that if spiritual transcendence was to be studied scientifically, it would be necessary to identify the basic phenomena that are common to all spiritually transcendent experiences. In *The Varieties of Religious Experience* (1902), he specifically identified four main characteristics of powerful spiritual experiences that he found to be common to individuals from all religions and that he summarized as follows:

- *Ineffability*: experiences that are so unique that they defy expression and cannot be adequately expressed through language
- *Noetic quality*: experiences that provide an insight into the nature of the universe that cannot be explained through the intellect
- *Transience*: experiences that tend to occur for a brief duration, although they often have long-lasting effects on the individual
- *Passivity*: experience of these events as a passive recipient, where one's own "will is in abeyance" and as if "grasped and held by a superior power" (James, 1902)

Over time, other scholars have offered similar models for mystical experiences, at times adding other characteristics to James' list but always including James' four primary attributes. For example, the philosopher Douglas Shrader (2008) acknowledged James' four characteristics of mystical experiences and added three more aspects: a sense of oneness, timelessness, and the feeling that one has encountered "the true self" beyond difference and duality. Shrader's three additional aspects of mystical experience fit well with our neuropsychological model as overlapping characteristics of selflessness, but noteworthy here is that James' core model has remained the standard for over 100 years.

We propose adding another core characteristic to James' list: a reduced sense of self. We argue that this diminished sense of self allows for (or may even be necessary for) what have been described as the unitive experiences associated with spiritual transcendence. That is, in order to experience unity with the divine, a profound connection to the cosmos, a special connectedness with nature, etc., one must attain a diminished sense of self.

3 UNIVERSAL FOUNDATIONS FOR SPIRITUAL EXPERIENCES

Like James, later philosophers such as Walter Stace and Aldous Huxley have also regarded the intense mystical and transcendent experiences of remarkable individuals recorded in a wide variety of religious faiths as comparable and having common psychological foundations. However, scholarly perspectives on the cross religious comparability of mystical experiences have remained mostly negative until only very recently. The ongoing problem for the dissenters, sometimes referred to as "contextualists," was most effectively stated by Steven T. Katz, Professor of Religion at Boston University, in a famous essay (1978), that regarded all attempts at generalizing and comparing mystical or transcendent experiences, whether religiously based or not, as unproductive because the cultural and social contexts in which they had occurred was vastly different. This is the famous and ongoing argument between "contextualists" and their antagonists, like James and others who have been termed "pluralists" and who maintain that mystical experiences are comparable cross-culturally and historically. For a time, it seemed that the contextualists had won the argument, but this outcome now appears to be changing. Perhaps, there is value in both positions, and the sway of these divisive arguments moves back and forth like a pendulum. Specifically, the pluralist viewpoint is gaining plausibility as spiritual experiences across different religions are being shown to have common neuropsychological foundations.

In part, recent developments in neuroscience are helping overcome contextualist concerns and suggest effectively that the cross-cultural comparisons of mystical experiences are not only possible but also extremely useful. A major contextualist complaint has focused on contentions about ineffability or the impossibility of accurately conveying or communicating mystical or spiritually transcendent experiences through language. Many who dispute the comparability of spiritually intense religious experiences, like Wayne Proudfoot, Professor of Religion at Columbia University, say that claims of ineffability are merely clever strategies used to protect such accounts from being discounted as unprovable and perhaps even completely imagined. This criticism, however, is better understood as part of a larger bias that has overprivileged language and at times has also been extended to explorations on the nature of consciousness, when it is argued that without language, conscious awareness simply cannot occur. But empirical evidence from neuroscientific research refutes the purportedly required connection between language and consciousness, in a significant reorientation that also has implications for the comparability of experiences of spiritual transcendence. An overreliance on language as the basis for conceptual thought and by implication for higher levels of awareness is no longer tenable. As the religious studies scholar of mysticism, Jason Blum has recently pointed out:

> Recent scientific evidence suggests not only that nondiscursive dimensions of consciousness can be identified but also that they constitute aspects of consciousness that appear to be more foundational to it than its cognitive, conceptual components. This indicates that the roles of emotion and language in experience—both mystic and otherwise—need to be fundamentally reconsidered. *(Blum, 2014, p. 156)*

Thus, it is possible that the capacity for human experiences of spiritual transcendence may have been hardwired in the brain before the development of human speech and language capabilities. What is becoming increasingly apparent is that the spiritually transcendent experiences of individuals from diverse faith traditions, cultures, and historic periods tend to share many of the same phenomenological characteristics.

4 INTENSE RELIGIOUS EXPERIENCES: PSYCHOPATHOLOGY OR MENTALLY HEALTHY?

An important question to ask is why expressions of profound religious and spiritual experience have often been regarded negatively, to the point where the spiritually transcendent experiences of religious mystics are often seen as evidence of psychological affliction? William James began the development of the fledgling field of the psychology of religion at the beginning of the 20th century based on his enthusiasm for studying the intense spiritual experiences of mystics, saints, and founders of religious faiths. However, despite his scientific outlook, attitudes toward intense mystical experiences remained mostly negative. The derogatory relationship between powerful spiritual experiences and mental health has persisted well into the late 20th century, with extreme mystical and religious convictions often seen as a form of psychopathology. In the early 20th century, the unchecked rise of positivism and the intense publicity of events like the Scopes trial, and the apparent defeat of religion by rational science seemed imminent, influencing developments within psychology as well.

In the latter half of the 20th century, an essential reference tool for psychiatrists and psychologists was developed, called the *Diagnostic and Statistical Manual (DSM)*, which became the central guide used to ascertain symptoms and prognoses for a wide variety of mental and psychological disorders. The scientific classification of psychiatric and psychological conditions has a complicated history that goes back well into the 19th century and cannot be recounted here, but the wide influence of the *DSM* cannot be overstated. It was originally developed as an offshoot of the US Army's desire to classify the medical status of service personnel and veteran outpatients. The *DSM's* classification of mental disorders was initially developed by borrowing 10 categories for psychoses and psychoneuroses and seven categories for disorders of character, behavior, and intelligence from the *International Classification of Diseases (ICD)* published by the World Health Organization in 1948 (American Psychiatric Association, 2013). The first edition of the *Diagnostic and Statistical Manual (DSM-I)* was prepared on this basis and released in 1952, published by the American Psychiatric Association (APA), the largest scientific and professional organization of psychiatrists in the world (founded in 1844).

Since the first edition of the *DSM* in 1952, there have been several subsequent and equally authoritative editions. Thus, in the third edition of the *DSM-III*, published in 1980 and revised in 1987 (*DSM-III-R*), many mystical and religious dispositions were still deemed to be pathological and a mental health hazard to those who exhibited them. There was no indication of the possible role of religious or spiritual factors in helping patients cope with mental illnesses. Most notably, it was only in 1994, with the publication of the *DSM-IV*, that "strong religious belief" that had previously always been officially classified as a form of mental disorder by the American Psychiatric Association, had its status reversed (see Dein, Cook, Powell, & Eagger, 2010 for a review). What prompted this sudden change?

One problem with prior diagnoses of mystical and religious experience as forms of psychological pathology was that they were based for the most part on perceived and assumed but unverified notions, influenced by a variety of social and cultural attitudes (not addressed here but a worthy topic for research). Expressions of intense religious or spiritual sentiment were considered as evidence of delusional behavior and psychological maladjustment. But somewhere along the way between the *DSM-III/ III-R* (1980/1987) and the publication of the *DSM-IV* in 1994, this attitude changed abruptly, as the psychological outlook on intense mystical and religious experiences was reevaluated. This newer perspective has continued in the latest edition, *DSM-5*, published in 2013 (American Psychiatric Association, 2013).

In regard to potent religious and spiritual experiences, there had only been limited empirical research on individuals deemed to be suffering from intense religious "delusions" and "hallucinations," many of whom might be better regarded as religious mystics. As more detailed psychiatric and psychological research accumulated, it was found that individuals who had had strong transcendent experiences were often psychologically healthier than matching controls who had not had intense mystical or religious experiences. If mystics were supposed to have confused or disordered minds, then why did they exhibit higher levels of mental health and psychological well-being (Newberg, d'Aquili, & Rause, 2001)? This led to a reevaluation and reformulation of psychiatric and psychological orientations toward mysticism. The history of the creation and development of the *DSM* show how contemporary psychological attitudes and understandings of intense spiritual experiences have changed considerably.

Part of the change in the conceptualization of mystical experiences can be attributed to a growing body of evidence showing that adherence to a particular religious faith is linked with better mental health. Individuals with religious involvement through belief and practice tend to show lower rates of drug abuse, alcoholism, divorce, and suicide than is found in the population at large (for a review, see Park et al., 2016). Studies have also shown that better physical health is associated with participation in organized religion (and even in just holding religious dispositions). Statistically, people who practice in mainstream faith traditions live longer and have fewer strokes, lower levels of heart disease, lower blood pressure, and improved immune function compared with the population at large (see Koenig, 2008).

5 COMPARING THE RELIGIOUS EXPERIENCES OF PSYCHOTICS AND MYSTICS

Granted, there are people who suffer from problematic religious hallucinations, for example, schizophrenics with religious delusions, which may include problems like an inflated sense of superiority or delusions of grandeur. Such individuals may correctly be termed psychotic, as they appear to have psychological or psychiatric disorders and do not exhibit better states of mental health. Notably, as was pointed out by Newberg, d'Aquili, and Rause, in their book *Why God Won't Go Away* (2001), psychotics and mystics can be clearly distinguished from one another psychologically, as they each respond to their intense spiritual and religious experiences in very different ways.

For example, in contrast to psychotics, mystics typically describe their spiritually transcendent experiences as ecstatic and joyful (okay the story of Job appears as an exception) and typically refer to feeling a sense of serenity, wholeness, and profound love from these experiences. Psychotics, in contrast, are often confused and terrified by their religious hallucinations, which may leave them in a severely distressed state and often highly fearful of a vengeful or angry God or higher power.

In many religions, there are specialized preparatory rituals for those seeking transformative mystical and transcendent experiences, which often involve purposeful separation from the everyday world as adherents seek a special and longed for spiritual experience. These special ritual preparations and the experiences they may precipitate are seen in a positive light, even if parts of the ritual preparation or even the profound spiritual experience itself might have been unpleasant or disturbing (see Otto (1917/1950), on the fascinating and terrifying aspects of religion). At the end of their period of separation from the community, when they have often entered temporarily into liminal (neither here or there) states, the mystic returns to the everyday world of their community and shares these experiences with others, often reintegrated into society with enhanced social status and position (Turner, 1969). In contrast, for the psychotic, such states are not deliberately induced, and their withdrawal from everyday life is involuntary and these experiences as distressing, and often result in driving such individuals into even deeper states of social isolation and disconnectedness.

Another significant difference is that mystics typically experience a high degree of sensory complexity that is coherent and multidimensional during their period of enhanced transcendence. In contrast, psychotics often tend to experience their hallucinations through a single sensory system, perhaps seeing a vision, hearing a disembodied voice, or feeling a presence, but these sensory experiences usually do not occur simultaneously. Further, while the mystic may have only a limited number of transcendent experiences in their lifetime and sometimes only one profound life-changing experience, individuals with psychotic religious hallucinations typically experience them on a much more regular basis, which often involve the same perceptual distortion or delusion each time. When the psychotic returns to everyday consciousness, they find their hallucinations to have been fragmentary, often dreamlike, and frequently interpret them as a result of their own mental dysfunction. In addition, their ability to clearly recollect such experiences often becomes severely diminished, whereas mystics typically remember their transcendent experiences with clarity and a strong sense of the reality of these occurrences. The mystic cannot be dissuaded from their sense that what they have experienced was genuine and their sense of the reality of such experiences does not dissipate over time, as it often does for the psychotic.

Psychotics in delusional states often express feelings of religious grandiosity and a strongly elevated sense of their own importance, perhaps expressed as a connection to God or higher force, and they may even come to see themselves as being a higher power. In contrast, mystics typically indicate that during a spiritually transcendent episode, they experience a loss of their ego, lessened pride and self-importance, and often a greatly reduced sense of self that may be expressed as merging with the infinite, the absolute, or God. While their acclaim may rise due to public recognition of their profound spiritual experiences, such individuals typically become humbled and show heightened modesty after these experiences and may even take on saintly qualities.

Thus, as William James understood long ago, we need to look closely at the lives of mystics and major religious figures for more complete insight into the nature of spiritually transcendent experiences. We suggest that such individuals and their spiritual experiences have a common neuropsychological ground related to the profound sense of the loss of the self that occurs, despite their embeddedness into a wide variety of historical periods, cultural contexts, and religious backgrounds.

6 SPIRITUAL TRANSCENDENCE AND THE SENSE OF SELF

It is reasonable to assume that the interpretations of spiritually transcendent experiences conveyed by individuals from different religious orientations and time periods might be unique to their particular faith tradition. However, it is difficult to understand how, for example, monotheistic spiritual experiences, found in Judaism, Christianity, or Islam, are somehow comparable with those from (so-called) polytheistic religions such as Hinduism or versions of Buddhism that are nontheistic. How could the emotional experience of the *unio mystica* (specifically the experience of unity with God) for Christians be similar to the experience of overcoming the false sense of the self (*anatman*) for Buddhists or to the nontheistic beliefs of avowed atheists? If there is a common foundation that can explain the similarities in these experiences across diverse faith traditions, it is likely to be related to common neuropsychological processes and the complex interactions of neurological networks.

Several researchers (Beauregard & O'Leary, 2007; McNamara, 2009; Newberg et al., 2001), including the current researchers (Johnstone & Glass, 2008; Johnstone, Bodling, Cohen, Christ, & Wegrzyn, 2012; Johnstone, Cohen, Konopacki, & Ghan, 2016), have suggested that the reported diminishing of the sense of the self experienced during spiritual transcendence is one neurophysiological/neuropsychological link that explains the similarity of such experiences across different faith traditions. Additionally, in descriptions of spiritual transcendence in the literature of many religious traditions from a wide variety of geographic regions and historical time periods, allusions to the loss of self are frequently found. Even the neuroscientist and avowed atheist Sam Harris acknowledges that expressions of the loss of sense of self in experiences of spirituality have been widespread throughout human cultures:

> There is no question that people have 'spiritual' experiences … Every culture has produced people who have gone off into caves for months or years and discovered that certain deliberate uses of attention—introspection, meditation, prayer—can radically transform a person's moment to moment perception of the world. I believe contemplative efforts of this sort have a lot to tell us about the nature of the mind … Selflessness is a quality of consciousness that can be subjectively discovered. *(Harris, 2007)*

Our premise is that selflessness can be explored both subjectively and objectively. But how do we explore the similarities between the spiritually transcendent experiences of individuals from specific religious traditions with the transcendent experiences of atheists, agnostics, skeptics, and human secularists? This question is perhaps increasingly relevant today as over one fourth of the US population now report being "spiritual but not religious" (Pew Research Center, 2017).

Our model of selflessness connects the religious and nonreligious experiences that involve transcendence with the loss of the sense of self (i.e., neuropsychological selflessness) and will help quantify such occurrences more accurately. Our approach also demarcates an important methodological separation between religion and spirituality. Specific religious activities, practices, and beliefs are identifiable, and one could argue that every religion has specific contextual and interpretive characteristics. For example, drawing on attribution theory from psychology, the religious studies scholar Ann Taves (2009) has argued that an experience only becomes a "religious experience" when it is attributed as such by those who have had that experience. But even if this were to constitute the definition of "religious experience," it does not pose a significant limitation to understanding the mechanisms of spiritual transcendence.

Whether a transcendent experience is identified as religious or not, attributional features are not the defining neuropsychological characteristic of such experiences. Experiences of spiritual transcendence are established on a neurophysiological level as variants of the loss of the sense of self, which at high levels can cause momentous changes in awareness. In terms of neuroscience, a spectrum or range of transcendent experiences that involve losing some degree of the sense of self can be described. A person may become immersed in music, in a book they are reading, in a poem or a work of art, while playing sports, when engaging in a creative activity or project, when seeing a beautiful landscape or sunset, etc. Each of us has transcendent experiences every day, when we become intensely focused on or lost in the task at hand. Our sense of time or space may fade away temporarily, until we return to our nontranscendent self-awareness and realize this has occurred. As Newberg et al. (2001) have pointed out, this phenomena can be expressed as a "unitary continuum" that goes from every day to mystical experiences. As they clearly expressed,

> The arc of this continuum links the most profound experience of the mystics with the smaller transcendent moments most of us experience every day, and shows that, in neurological terms, the two are different essentially by degree. (*Newberg et al., 2001, p. 115*)

When transcendent experiences of selflessness are powerful enough, they can leave a lasting emotional, transformative, and life-changing impact on a person and at times even on others who may later come to learn about such profound occurrences of spiritual transcendence.

7 THE PSYCHOLOGY OF SPIRITUAL TRANSCENDENCE

Religious practices may indicate the importance of attaining some level of transcendence in everyday practices, through prayer or meditation, for example, in order to connect with higher powers or attain greater awareness. But recent psychological research has increasingly noted that it is a greatly heightened loss of the sense of self that is emphasized during profound mystical experiences. For example, the psychologist Ralph Hood has pointed out the following:

> It is well established that an experience of unity is the defining feature, not of selfhood, but of self-loss in mystical experience. *(Hood, 2002, p. 2)*

Hood makes an interesting distinction between "the unity of self found in reflection" and "the unity of self devoid of attributes" in attempting to clarify the different potentials of spiritual transcendence. He says,

> Our preference is to call the former self the reflexive self ... and the latter self the transcendent self ... To lay our cards down on the table, we suggest that the unity of the reflexive self is what is lost in mysticism, whereas the unity of the transcendent self is found. *(Hood, 2002, p. 4)*

In other words, in profound cases of spiritual transcendence, the regular sense of self loses its qualities or regular characteristics completely, and this defines a transcendent mystical experience.

Foreshadowing this perspective, the famed psychologist Abraham Maslow, the developer of transpersonal psychology, said this about spiritual transcendence:

> Transcendence refers to the very highest and most inclusive or holistic levels of human consciousness, behaving and relating, as ends rather than means, to oneself, to significant others, to human beings in general, to other species, to nature, and to the cosmos. *(Maslow, 1971, p. 269, also quoted in Garcia-Romeu, 2010, p. 26)*

Thus, during experiences of transcendence one not only gets lost in oneself but also often connects or becomes unified with something larger that the self. Accordingly, it may be said that self-transcendence expresses the highest level of consciousness if Maslow is correct.

Importantly, recent research by Koltko-Rivera (2006) has revealed that the conventional psychological description of Maslow's hierarchy of needs based on an end point of "self-actualization" has been overwrought. The problem is that this psychological outlook does not take into account Maslow's later thought where he amends his model of self-actualization, which he had previously regarded as the ultimate psychological state achievable, and places self-transcendence as the definitive experience that is beyond self-actualization, which he had originally touted in his transpersonal psychology. As Maslow said,

> The fully developed (and very fortunate) human being working under the best conditions tends to be motivated by values which transcend his self. They are not selfish anymore in the old sense of that term. Beauty is not within one's skin nor is justice or order. One can hardly class these desires as selfish in the sense that my desire for food might be. My satisfaction with achieving or allowing justice is not within my own skin ... It is equally outside and inside: therefore, it has transcended the geographical limitations of the self. Thus one begins to talk about transhumanistic psychology. *(Maslow, quoted in Koltko-Rivera, 2006, p. 306)*

To clarify his new position, Maslow says the following:

> Let me call behavioristic psychology the first psychology, Freudian psychology the 2nd, and humanistic psychology the 3rd. And then what I think of as a real possibility, what I am fascinated with, is the psychology of transcendence & of ends, the transhuman or transpersonal (really should be "transcendent") psychology = the 4th psychology. *(Maslow, quoted in Koltko-Rivera, 2006, p. 306)*

Thus, it is self-transcendence that is the highest form of psychological attainment according to Maslow.

Similarly, the psychologist Viktor Frankl also downplayed the importance of self-actualization, and he too connected it with the wider goal of self-transcendence. For Frankl, as for Maslow, self-transcendence represented the ultimate psychological achievement:

> By declaring that man is responsible and must actualize the potential meaning of his life, I wish to stress that the true meaning of life is to be discovered in the world rather than within man or his own psyche, as though it were a closed system. I have termed this constitutive characteristic "the self-transcendence of human existence." It denotes the fact that being human always points, and is directed, to something or someone, other than oneself—be it a meaning to fulfill or another human being to encounter. The more one forgets himself—by giving himself to a cause to serve or another person to love—the more human he is and the more he actualizes himself. What is called self-actualization is not an attainable aim at all, for the simple reason that the more one would strive for it, the more he would miss it. In other words, self-actualization is possible only as a side-effect of self-transcendence. *(Frankl, 1985, p. 133)*

Thus, for a considerable time, psychologists have been recognizing the importance of striving to attain experiences of selflessness through self-transcendence, in order to attain the highest levels of psychological well-being, fulfillment in life, and the highest levels of awareness.

8 SPIRITUAL TRANSCENDENCE AS A PERSONALITY TRAIT

One of the problems inherent in studying spiritual transcendence is that it has not been clear how to conceptualize it methodologically, therefore making it hard to measure. For example, it is often not clear if spiritual transcendence involves cognitive or affective processing or if spiritual transcendence is perhaps a completely unique process. In fact, recent psychological research is suggesting that the propensity to have spiritually transcendent experiences may be best considered as a distinct personality trait. Personality characteristics have traditionally been conceptualized as those psychological and behavioral traits that are inheritable, unique to each individual, and generally stable over one's life. At the current time, personality is primarily conceptualized according to the *Five Factor Model* (FFM; Costa & McCrae, 1992) that suggests that all individuals possess varying degrees of five general personality traits including extraversion, neuroticism, agreeableness, openness to experience, and conscientiousness. Just as individuals are genetically predisposed to have certain physical characteristics (e.g., height, weight, and hair/eye color), they are also inclined to exhibit certain personality characteristics. One only has to think of the similar personality characteristics of parents and their children to understand that personality traits are indeed inherited.

In addition to these five commonly accepted personality traits, the psychologist Ralph Piedmont has proposed that spiritual transcendence may be best conceptualized as an additional personality trait. Piedmont says,

> Spiritual Transcendence represents a broad-based motivational domain of comparable breadth to those constructs contained in the FFM and ought to be considered a potential sixth major dimension of personality. *(Piedmont, 1999, p. 985)*

This suggests that in addition to responding to the world in psychologically consistent manners (e.g., as extraverted and agreeable), individuals are also predisposed to experience the divine, nature, or the cosmos with a spiritually consistent orientation. Piedmont describes spiritual transcendence as follows:

> Spiritual Transcendence refers to the capacity of individuals to stand outside of their immediate sense of time and place to view life from a larger, more objective perspective. This transcendent perspective is one in which a person sees a fundamental unity underlying the diverse strivings of nature ... *(Piedmont, 1999, p. 988)*.

Furthermore, he argues that while some individuals are predisposed to experience their spiritual side more deeply, others are perhaps less so inclined.

To support his theory, Piedmont conducted several studies showing that spiritual transcendence is, in fact, consistent as a separate personality trait. Using the *Spiritual Transcendence Scale* (Piedmont, 1999) along with a standard measure of personality traits, he was able to show that spiritual transcendence operated as a unique construct that was distinct from the other five traditional personality traits. He also found that spiritual transcendence was identifiable as a unique construct based both on self-rating and on the ratings of others, meaning people could consistently describe their own spiritual predisposition and so could others who knew them. Reinforcing this view of spiritual transcendence, Piedmont and Leach (2002) and Piedmont (2007) found, through research using the same methodology with individuals in India and the Philippines, that it also appears to be a distinctive personality trait cross-culturally.

Notably, Piedmont also distinguished religiousness from spiritual transcendence, and he indicated that spiritual transcendence is a larger category than religion:

> Spiritual Transcendence is clearly different from religiousness ... the former emphasizes a personal search for connection with a larger sacredness while the latter provides a more social emphasis on encountering the divine. Although ... [others] argue spirituality is one component of religiousness, I believe that Transcendence is much larger than religiousness. It represents a broad domain of motivations that underlie strivings in both secular and religious contexts, and this is what distinguishes it from other similarly named constructs ... *(Piedmont, 1999, p. 989)*

We agree with this assessment and find that it reinforces our sense that spiritual transcendence is a phenomenon that occurs cross-culturally and can be examined and compared across different religious (and nonreligious) traditions and orientations.

Other studies have also indicated that spiritual transcendence is also consistent with a personality characteristic given its general stability over time. Specifically, Brändström et al. (1998) showed that a person's spiritual propensity toward connecting with the divine was more stable over their lifetime than other personality characteristics. This is consistent with what would be expected for genetically inherited traits, and it suggests that it may be possible to empirically measure spiritual transcendence in scientific studies.

9 GENETIC STUDIES OF SPIRITUALITY

There is evidence that spirituality may be an inheritable characteristic. Studies of twins are a way to explore the genetic inheritability of specific behavioral traits and physical characteristics. Twin studies have consistently indicated that the tendency to be "spiritual" appears to be genetically inherited.

Identical twins develop from the same fertilized ovum (i.e., monozygotic twins) and have 100% of the same genes, while fraternal twins that develop from different ovum (i.e., dizygotic twins) have only 50% of the same genes (as do all siblings). As a result, identical twins have the same general physical and psychological characteristics (i.e., they look alike and act similarly) because they have developed with identical genetics. In contrast, fraternal twins will have less similar physical and psychological characteristics given they share only 50% of the same DNA. By studying similarities and differences in traits and behaviors of identical and fraternal twins, it is possible to explore the degree to which these characteristics appear genetically inherited, especially when environmental factors are consistent. Theoretically, if a characteristic is genetically inherited, then identical twins should show twice the concordance rate or correlation of expression for that trait than do fraternal twins, given that identical twins have twice the number of similar genes.

Researchers in Australia have established the most comprehensive database of twins to date through the Australian National Health and Medical Research Council (NHMRC) Twin Registry. They have collected comprehensive data on over 25,000 pairs of twins, developing an excellent platform for examining genetic and environmental influences on a variety of traits and characteristics. In one study from this database, researchers compared "spiritual transcendence" for over 3000 sets of twins, aged 50 years and older (Kirk, Eaves, & Martin, 1999). In this twin study, they found that the correlation between identical male twins in terms of spiritual transcendence was 0.40, while for fraternal male twins it was only 0.18. Similarly, for female identical twins, the correlation was 0.49, but only 0.26 for female fraternal twins. These fairly consistent results suggest a genetic factor operating here for the disposition to spiritual transcendence. Interestingly, it was also found that orientations toward spiritual transcendence among twins did not correlate with church attendance, suggesting that religious participation did not influence spiritual disposition.

Although these results are thought provoking in terms of the fairly even ratios of differences found between identical and fraternal twins in terms of the propensity for spiritual transcendence, it is possible that the similarities between the twins in terms of spirituality may have been primarily related to environmental factors such as growing up in the same environments and households.

Although it is extremely rare for twins to be raised apart, a group of researchers in Minnesota had the unique opportunity to compare the genetic and environmental influences on the spirituality of twins who were separated at birth. These researchers tracked down 84 pairs of twins (53 identical and 31 fraternal) who were adopted at birth and raised apart (Bouchard, Lykken, McGue, Segal, and Tellegen, 1990; Bouchard, McGue, Lykken, and Tellegen, 1999). The Minnesota researchers investigated the differences among all the twins raised apart in terms of five different aspects of "religiousness," and the results indicated that all five aspects were genetically inherited. Bouchard, McGue, Lykken, and Tellegen (1999) also investigated relationships between twins who were raised apart in terms of their intrinsic religiousness (i.e., spiritual orientation) and their extrinsic religiousness (i.e., culturally influenced religious beliefs and behaviors). In this study, spirituality was considered as a psychological attribute, while religiousness was regarded as a culturally influenced behavior. It was found that spirituality (intrinsic religiousness) was genetically inheritable with a correlation of 0.37 between identical twins but only 0.20 between fraternal twins. In contrast, the results regarding culturally influenced religious beliefs and behaviors (i.e., extrinsic religiousness) were found to be "ambiguous" and showed no detectable patterns. Thus, it appears that the inclination toward spiritual transcendence, expressed as a feeling of closeness to the divine, may have a genetic component.

If our ability to be spiritually oriented is in part genetically inherited, this implies that our capacity to have experiences of spiritual transcendence may involve a certain degree of neurological hardwiring. It is commonly acknowledged that our sensations, emotions, thoughts, and behaviors are based in the brain. If the propensity for spirituality is genetically inherited, similar to cognitive ability or personality characteristics, then it too may be related to certain neural processes and networks.

10 MEASURING SPIRITUAL TRANSCENDENCE

The previous sections indicate that "spiritual transcendence" can be effectively identified as a scientifically valid construct. Specifically, it can be succinctly defined by common phenomenological characteristics. Additionally, it has been shown to have unique characteristics as a personality trait that is stable over time and can be differentiated from other personality characteristics. Finally, it has been shown to be a trait that could possibly be genetically inheritable. However, just as developing concise definitions of spiritual transcendence has proved to be difficult, attempts to quantify it empirically have been equally problematic.

Several measures of spiritual transcendence have been developed and used in research over the past several decades, including the *Self-Transcendence Scale* (Reed, 1991), the *Spiritual Transcendence Scale* (Piedmont, 1999), the *Adult Self-Transcendence Inventory* (Levenson, Jennings, Aldwin, & Shiraishi, 2005), and the *Self-Transcendence Scale* component of the *Temperament and Character Inventory* (TCI; Cloninger, Pryzbeck, Svrakic,

TABLE 1 Examples of Items Regarding Selflessness and Unitive Experiences From Self-Transcendence/Spirituality Scales

---

TCI Self-Transcendence Scale (Cloninger, Pryzbeck, Svrakic, & Wetzel, 1994)
- Sometimes, I have felt as if I was part of something with no limits or boundaries in time or space
- I sometimes feel so connected to nature that everything seems to be part of one living organism

Self-Transcendence Scale (Piedmont, 1999)
- I have had a spiritual experience where I lost track of where I was or the passage of time
- All life is interconnected

Self-Transcendence Scale (Reed, 1991)
- I experience a deep communion with God
- I find meaning in my spiritual beliefs

Adult Self-Transcendence Inventory (Levenson, Jennings, Aldwin, & Shiraishi, 2005)
- I feel part of something greater than myself
- I often have a sense of oneness with nature
- I feel that my individual life is a part of a greater whole
- My sense of self has decreased as I have gotten older

Expressions of Spirituality Inventory-Revised (MacDonald, 2000)
- I have had an experience in which I seemed to transcend space and time
- I have had an experience in which I seemed to be deeply connected with everything
- I have had an experience in which I seemed to go beyond my normal everyday sense of self
- I have had an experience in which I seemed to merge with a power of force greater than myself

Inspirit (Kass, Friedman, Lesserman, Zutterman, & Benson, 1991)
- How often have you felt as though you were very close to a powerful spiritual force that seemed to lift you outside of yourself?
- Indicate whether you agree or disagree with this statement: God dwells within you.
- How close do you feel to God?
- Have you ever had this experience:
  ○ An experience of God's energy or presence
  ○ A feeling of unity with the earth and all living things

Brief Multidimensional Measure of Religiousness/Spirituality (Fetzer Institute & National Institute on Aging Working Group, 1999)
- I think about how my life is part of a larger spiritual force
- I experience a connection to all of life
- I desire to be closer to or in union with God
- I feel God's presence

& Wetzel, 1994). To help demonstrate how spiritual transcendence is conceptualized and measured by these scales, representative items from several of these measures are presented in Table 1.

It is important to note that although these different measures share the same designation indicating that they measure transcendence, they differ in the ways in which they conceptualize specific dimensions of transcendence. However, it appears that many of the items they use to measure reflect respondents' increased sense of connection to the universe or the divine are coupled with an associated experience of a decreased sense of the self. For example, according to Cloninger,

> Self-transcendence refers generally to identification with everything conceived as essential and consequential parts of a unified whole. This involves a state of "unitive consciousness" in which everything is part of one totality. In unitive consciousness, there is no individual self because there is no meaningful distinction between self and other—the person is simply aware of being an integral part of the evolution of the cosmos. This unitive perspective may be described as acceptance, identification, or spiritual union with nature and its source. (*Cloninger, Svrakic, & Przybeck, 1993, p. 981; they cite Underhill, 1911*)

Notably, the *Self-Transcendence Scale* developed by Cloninger and his colleagues measures three distinct aspects of spiritual transcendence, using subscales that include *self-forgetting, transpersonal identification*, and *spiritual acceptance*. The *self-forgetting* subscale measures the tendency of individuals to lose themselves in thought and experience. The *transpersonal identification* subscale assesses one's ability to feel connected with humanity, nature, and the universe. The *spiritual acceptance* subscale measures belief in forces that cannot be rationally comprehended or objectively proved.

If we compare Cloninger's and Piedmont's *Self-Transcendence Scales*, we see that both measures include many similar questions, although Piedmont's scale conceptualizes spiritual transcendence using three constructs that are different than Cloninger's. Piedmont explores *connectedness*, which measures the belief that one is part of a larger human community; *universality*, which measures one's belief in the unitive nature of life; and, third, *prayer fulfillment*, which measures feelings of joy and contentment that result from personal encounters with a transcendent reality. Overall, the items of each scale are generally similar, suggesting that their tests are theoretically sound and have face validity, meaning that they measure a common construct that can be labeled as "spiritual transcendence." However, in order to use them in scientific studies, it is also necessary to show that they have psychometric validity, and accordingly, several studies have found that "spiritual transcendence" is a psychometrically sound construct (Garcia-Romeu, 2010).

Cloninger's *Self-Transcendence Scale* has been the most thoroughly investigated of these measures of spiritual transcendence to date, particularly in quantitative studies (see Garcia-Romeu, 2010, for a review). A variety of studies has shown that his *self-transcendence scale* produces similar and reliable results over time and that it measures the same general construct as the other indicated measures of spiritual transcendence, which is called evidence of "convergent validity." Additionally, it is distinguishable from other measures that use dissimilar constructs, which means it also achieves "divergent validity." Thus, the concept of spiritual transcendence and its measures appears to be both theoretically and scientifically valid from a psychometric perspective. This suggests that measures of spiritual transcendence have a common meaning across studies and can be evaluated using scientific methods and compared across different studies.

11 NEED FOR NEUROSCIENTIFIC MODELS OF SPIRITUAL TRANSCENDENCE

The information presented in this chapter suggests that spiritual transcendence is a unique experience that can be defined phenomenologically, measured scientifically, and compared statistically. We define "spiritual transcendence" as a sense of increased cosmic unity or emotional connection with the divine or the infinite, conceptualized according to an individual's worldview (e.g., belief in God, in multiple aspects of the divine, special connections with nature or the universe, and experiencing the existential void). Of primary importance, we focus on the reduced sense of self that is commonly reported as a core characteristic of spiritual transcendence. We believe this perspective is critical given that the "self" continues to be explored from the perspectives of both the neurosciences and humanities and can serve as a bridge between these diverse disciplines. As such, references to "spiritual transcendence" in the following chapters refer to the relationship between reducing the sense of self and the experience of connecting with the divine, whether conceived as connecting with something larger than oneself or as an experience of complete emptiness; they all involve a greatly diminished sense of the self.

Although initial scientific studies have focused on spiritual transcendence as a personality trait, more recent research is beginning to focus on the neurological foundations of spiritual transcendence. If the propensity for experiences of transcendence is in part genetically inheritable, then it is also conceivable that it is also neurophysiologically hardwired. This suggests that our ability to have experiences of spiritual transcendence is similar to the manner by which cognitive processes (e.g., memory, attention, and language) and personality traits (e.g., extraversion and agreeableness) are neurally based. If this is the case, then a model that can identify specific neuropsychological capabilities and associated neuroanatomical regions and networks that serve as universal foundations for spiritual transcendence is needed. But rather than assume that there is a specific "spiritual" neuropsychological process for spiritual transcendence, reflective of earlier problematic pursuits to find the neurological correlates of religion or to search for a God spot in the brain, what is needed is ways to determine the neuropsychological processes that provide the best explanation for experiences of spiritual transcendence.

We believe that the humanities have already provided strong indications as to the primary neuropsychological foundations for spiritual transcendence. The term "transcendence" offers some guidance, as it implies "the ability to move beyond." When describing spiritual transcendence, it is often said that it is the ego, the self, or the individual identity that is being transcended in order to attain spiritual connection with the divine or the cosmos. The texts and scriptures of many religious faiths support this outlook and routinely discuss the need to minimize the self, obliterate the self, move beyond the self, etc., in order to connect with higher powers and find meaning in life. Mystical experiences of spiritual transcendence are repeatedly described in terms of a diminished sense of self, achieved by making a connection to something larger than the self.

Over the past 25 years, neuroscientists have studied the manner by which the brain processes a sense of the self. Although the self has also been conceptualized in philosophical understandings, both in terms of being an eternal subject and permanent entity or occurring as a phenomenal object that is experienced consciously, scientific research is beginning to suggest that self is actually an integration of several neuropsychological processes that create a modulated *sense of self*. By conceiving of the self in this manner, it becomes possible to understand how the sense of self can be enhanced or diminished depending on context, circumstance, and behavioral practices. We believe that the humanities clearly suggest that the diminished sense of the self is a core foundation for spiritual transcendence. In order to develop an introductory neuropsychological model of the processes of spiritual transcendence, it is first necessary to provide evidence that the experience of the self is based on a range of interacting neuropsychological processes.

Recent neurological research connects the loss of the sense of self with experiences of unity and spiritual transcendence. The following section of this book will review neurological and neuropsychological studies and investigations to help develop the model for this interconnection by presenting case studies of individuals with neurological injury, primarily of the right hemisphere/right parietal lobe, who have experienced a distorted sense of self. It will also review empirical studies that show that the right hemisphere in general and right parietal lobe specifically process information that is specific to the self.

THE "SELF" AND SELFLESSNESS

Disorders of the Self

Selflessness is a quality of consciousness that can be subjectively discovered.
(Harris, 2007)

One of the more interesting methods used to understand how the brain processes information related to the self is to study individuals with unusual and very focal brain disorders. It has been argued that it is only by looking at the "abnormal" brain that we can truly understand how the normal brain functions. Several popular books describe such unique cases, including the *Man Who Mistook his Wife for a Hat* (Sacks, 1985); *Patient H.M.: A Story of Memory, Madness, and Family Secrets* (Dittrich, 2016);

Neuroscience, Selflessness, and Spiritual Experience
https://doi.org/10.1016/B978-0-08-102218-4.00003-1

43

© 2019 Elsevier Inc. All rights reserved.

An Odd Kind of Fame: Stories of Phineas Gage (MacMillan, 2002); and *Dueling Neurosurgeons* (Kean, 2015). These books tend to be a bit more interesting than more sterile academic articles and provide a novel way to learn how the brain works. Furthermore, how can anyone *not* be interested in reading about a man who mistook his wife for a hat?

In this chapter, we discuss unique neurological cases that have improved our understanding of the neuropsychological foundations of memory, language, and executive skills and in particular how these cases help us better understand how the brain processes information related to creating a "self."

1 "NO LONGER GAGE"

During the first half of the 19th century, brain functions were often studied through the "science" of phrenology. It was believed that different cognitive abilities and personality traits could be best inferred by evaluating the bumps on one's head (see Fig. 1). Although phrenology persisted for some time, knowledge regarding relationships between the frontal lobes and behaviors unexpectedly advanced in 1848 due to the misfortunes of a railroad worker who happened to be at the wrong place at the wrong time.

Phineas Gage was 25 years old when he incurred a very serious, focal brain injury while working on the railroad. He was tamping explosives when an unexpected explosion resulted in a three and one-half foot iron rod (i.e., basically a metal javelin) exploding up through his jaw and left eye orbit, passing through his left frontal lobe, and out the top of his head. Miraculously, he never lost consciousness, but shortly following his injury, it was reported that "Mr. G. got up and vomited; the effort of vomiting passed out about half a teacupful of the brain, which fell upon the floor" (Bigelow, 1850; as cited in MacMillan, 2002, p. 448).

Needless to say, it was a serious injury that left Mr. Gage blind in the left eye for the rest of his life. He reportedly recovered well from an intellectual standpoint, which was astounding to his peers and physicians, especially given the serious nature of his injury. Prior to his injury, Mr. Gage was described as "… a most efficient and capable foreman … a shrewd, smart business man, very energetic and persistent in executing all his plans of operation" (Harlow, 1868; as cited in MacMillan, 2002, p. 90). However, following his injury, Mr. Gage's friends, family, and coworkers all noticed him to exhibit atypical behaviors and significant changes in his personality. After his injury, he was described by his physician as follows:

> *The equilibrium or balance, so to speak, between his intellectual faculties and animal propensities, seems to have been destroyed. He is fitful, irreverent, indulging at times in the grossest profanity (which was not previously his custom),*

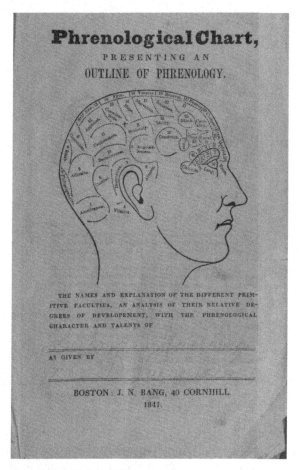

FIG. 1 Phrenology pamphlet from 1841.

manifesting but little deference for his fellows, impatient of restraint or advice when it conflicts with his desires, at times pertinaciously obstinate, yet capricious and vacillating, devising many plans of future operations, which are no sooner arranged than they are abandoned in turn for others appearing more feasible. A child in his intellectual capacity and manifestations, he has the animal passions of a strong man ... In this regard his mind was radically changed, so decidedly that his friends and acquaintants said he was "no longer Gage". **(Harlow, 1868; as cited in MacMillan, 2002, p. 92–93)**

This case made it increasingly clear that individuals are prone to be impulsive by nature but that the frontal lobes "inhibit" these impulses. It is now routinely understood by rehabilitation professionals that individuals with acute frontal lobe injuries can be impulsive/disinhibited and

II. THE "SELF" AND SELFLESSNESS

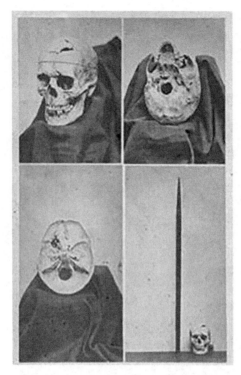

FIG. 2 Phineas Gage's skull. *Source: García-Molina, A. (2012). Phineas Gage and the enigma of the prefrontal cortex (Phineas Gage y el enigma del córtex prefrontal), Neurología (English ed.), Vol. 27, Issue 6, pp. 370–375. Copyright Elsevier.*

often say whatever is on their mind (e.g., they will tell you in no uncertain terms what they think of you when they do not want to do any more rehabilitation) or act on their first impulse in a given situation (e.g., help themselves to the food on their roommate's plate). Mr. Gage was one of the first case studies that clearly indicated that the frontal lobes and not the location of bumps on the head as phrenology had asserted are related to the regulation of behaviors and emotions and intrinsically linked to one's personality (Fig. 2).

2 Tan Tan

About the same time as Mr. Gage's case helped to clarify the role of the frontal lobes in inhibiting impulses, in the mid-1800s, a series of unusual cases in France helped to clarify the neurological foundations of expressive language. Mr. Leborgne (who is better known as "Tan Tan" for reasons that will become clear) was a young man in Paris who was hospitalized for intractable seizures. At age 30, he suddenly lost the ability to speak, other than to say "Tan Tan" (thus his nickname). He was noted to have generally intact cognitive abilities in all other areas, including the ability to comprehend everything that was stated to him. After approximately 20 years, he developed paralysis of the right arm and leg and was eventually treated by Dr. Paul Broca who had interests in the neurological origins of language. After Tan Tan's death, Dr. Broca conducted an autopsy and found a large lesion in the posterior inferior frontal gyrus, which he concluded was related to Tan Tan's specific and significant weaknesses in expressing himself.

Not long after, Dr. Broca treated another patient, an 84-year-old man named Lazare Lelong. He had dementia and ultimately lost the ability to speak, except for five words. Upon his death, an autopsy indicated a lesion in essentially the same region as found in Tan Tan (i.e., the posterior inferior frontal gyrus). Dr. Broca used these cases to conclude that expressive language abilities could be associated with a specific cerebral area. Since that time, the region for expressive language located in the left frontal lobe has been universally recognized as Broca's area.

3 THE MAN WITHOUT A MEMORY

Whereas the neurological foundations of language and inhibitory control were beginning to be understood as early as the mid-19th century, the neurological foundations of memory were not well understood until the 1950s. The most eminent memory researcher at that time, Karl Lashley, conducted extensive memory research with rats and concluded that memory was related to the entire brain and could not be associated with any specific brain region. However, the case of Henry Gustav Molaison (known as Patient HM until his identity was published after his death in 2008) radically changed the understanding of the nature of memory.

Mr. Molaison is thought to have incurred a brain injury at age 7 when he was struck by a car when riding his bike. He developed seizures at age 10 that became intractable, developing into a generalized epilepsy that involved his entire brain. As is common for individuals with uncontrollable seizures, Mr. Molaison was eventually unable to work or live independently. At age 27, he agreed to have neurosurgery to remove his bilateral medial temporal lobes (i.e., primarily involving his hippocampus), which was determined to be the brain region where his seizures originated. It was thought that by removing the area where the seizures originated, his seizures would hopefully be eliminated or at least reduced.

At that time, the neuropsychological abilities specifically associated with the hippocampus, located in the limbic area of the brain, were generally unknown, but the impairments Mr. Molaison suffered following his surgery made this relationship perfectly clear. Generally, he did retain intact neuropsychological abilities in his cognitive and communication capacities (e.g., intelligence, language, and problem solving) following surgery. However, as he recovered, it became apparent that he had lost the ability to encode new memories on a moment-to-moment basis (called "anterograde memory"). He could generally recall information that he had learned during his childhood and up to 3 years prior to his surgery, including episodic memories related to events and people in his life that had occurred up until that time (referred to as "retrograde memory"). However, from the time of the surgery forward, Mr. Molaison was virtually unable to remember any new information (with a few exceptions) for more than several seconds. In essence, he became a man without the ability to create new memories. He could not remember if he had eaten breakfast 10 minutes after doing so and would need to be repeatedly reintroduced to individuals he had just met, even after meeting them on a regular basis. Mr. Molaison became that hypothetical person that others occasionally wonder about, a man without a memory. Although a very unfortunate situation for Mr. Molaison, his circumstances greatly improved our understanding of the neurological foundations of different memory processes.

4 BRAIN DISORDERS AND THE "SELF"

In the same way that these earlier case studies delineated relationships between cerebral regions and specific neuropsychological abilities (i.e., inhibition, speech, and memory), more recent neurological case studies demonstrate the manner by which specific brain regions are associated with the process of creating and perceiving the "sense of self." To understand how the brain creates a "self," consider the following.

Write down who "you" are. More than likely, you will describe your physical characteristics (e.g., age, size, and hair/eye color) and your psychological attributes (e.g., personality traits and interests). Furthermore, you will likely describe yourself in terms of your previous histories (e.g., the son/daughter of your parents, graduate of a specific school, and employee of such and such a company). You are "you," period. Correct? Well, maybe not. To challenge your conceptualization of what your "self" is, consider the following.

Raise your left arm. Pinch your left thigh. Run your left hand through the hair on the left side of your head. Are your left arm, left thigh, and left side of your head part of your "self?" Of course they are, right? It is a core assumption that our bodies are the physical foundation for our "selves." Could you ever conceive of a situation in which you could look at the left side of your body and deny that it is part of "you"? Is it possible that our sense of a physical self is actually based on neuropsychological processes in the brain and a change in those brain regions leads to a change in the neuropsychological process of constructing a self? If this is the case, it creates a highly unusual understanding of our conceptualization of a "self." It is easy to understand what happens to individuals if they experience significant impairments in memory (e.g., Henry Molaison), disinhibition (e.g., Phineas Gage), or language (e.g., Tan Tan and Lazare Lelong). However, try to imagine the characteristics of individuals who lost the ability to create a self or who no longer have the sense of "I-me-mine."

Numerous case studies throughout history have, in fact, suggested that certain parts of the brain are associated with specific neuropsychological processes that create a sense of the self, including a physical sense of self (i.e., a sense that this body is "mine"), a psychological sense of self (i.e., a sense that one's thoughts and emotions are "mine"), and an autobiographical self (i.e., a sense that recollections of past experiences and histories are "mine"). However, if this is the case, it begs the question of what happens if the sense of "mineness" is minimized or suspended.

5 LOSING THE LEFT SIDE

One of the most baffling symptoms associated with disorders of the right parietal lobe is a condition known as left-sided neglect (also known as unilateral neglect) in which individuals are unaware of the left side of their bodies, even though they have intact vision. Such individuals are generally unaware of the relationship of their body to objects in their left side of space. Like most neurological conditions, left-sided neglect can be experienced in many different ways and appear as a continuum (i.e., from mild to severe neglect). One of the most commonly reported symptoms for those suffering from left-sided neglect is the tendency to bump into things on the left side of space, such as doorjambs. All of us navigate our homes and surroundings without much conscious thought. We see a door, orient ourselves, and walk through it. Individuals with right parietal lobe-related left-sided neglect see a door, orient themselves to it (but with reduced awareness of the left side of the world), and then bump into the doorjamb on their left side as it is outside their field of spatial awareness.

Left-sided neglect can also be experienced in many other ways. For example, the brother of one of the authors (BJ) was 27 years old when he was driving on the road with his wife to Grant's farm in St. Louis. She noticed that her husband began to swerve to the left and drive over the yellow median lines and into the oncoming traffic and that he was apparently unaware that he was doing this. When they arrived and he was turning left into a parking lot, he cut the corner way too closely and almost hit a car coming out of the lot. His wife was completely baffled as to why he could not recognize that he was about to harm or even kill them with his careless driving—how could he not see that he was driving into oncoming traffic?

In addition, he was puzzled by her insistence that his driving was dangerous. Inside the park at Grant's farm, she noticed that he kept bumping into picnic tables on his left side. She did not notice any other unusual symptoms (e.g., aphasia and forgetfulness), and she did not know what to make of his problems. She called her mother-in-law (who was a nurse) for guidance, and after speaking with her, she then immediately took him to the hospital where he was found to be suffering a stroke in the right parietal lobe.

Think of the last time that you drove a car and inadvertently drifted to the left (e.g., reading a text, changing the radio station, and drifting off into a distracting daydream). There is a good chance that you quickly recognized that your car was veering left and immediately corrected your car's position. However, with the dysfunction of the right parietal lobe, you may not be aware that you are neglecting the left side of space and therefore may not be able to make a correction—after all, why correct for something of which you are unaware? These cases illustrate how our brain processes information about the relationship between our physical body (our physical "self") and the space around us.

Consider another case. At the University of Washington, a nurse in her mid-30s was hospitalized for a mild stroke in the right parietal lobe. Upon completing a neuropsychological evaluation, the nurse obtained a relatively low score on a reading test, which did not make sense given that she did not report any such problems. However, review of the actual reading form indicated that the vast majority of her errors occurred when she was reading the first word on the left side of each row (these reading tests involve reading rows of words, left to right, top to bottom, and easiest to most difficult). She had neglected to read the first word in each row of the test. Although her vision was entirely intact, she could not fully orient herself to the left side of the test form (i.e., she neglected it and ignored it). After her errors were noted, the technician asked her to reread the list of words, this time having the first word in each row pointed out to her. When the technician reoriented her to the left side of the test, she was able to read the majority of words without difficulty.

After the nature of her difficulties became clear, it was concluded that she did not have a reading disorder, but rather she had a very subtle form of left-sided neglect that only affected the far left side of her perceptual world. Upon questioning, she also admitted to occasionally bumping into things on her left side. Based on these findings, it was determined that it was best for her to not return to her work as a nurse. No one wants to work with a nurse if she ignores the far left side of space, which could include neglecting to pay attention to the dosages of medications she dispenses (e.g., administering 50 mg of a drug when 250 mg is called for).

II. THE "SELF" AND SELFLESSNESS

6 CLOCK DRAWING

There are several simple tests health professionals use to determine if individuals are experiencing left-sided neglect associated with right parietal lobe dysfunction. One of these tests involves being shown a circle, which represents the face of a clock. Individuals are asked to place the numbers of the clock (i.e., 1–12) within the circle and then place the short and long hands of the clock to indicate a specific time (e.g., 11:10). Neurologically intact individuals have little if any difficulties with this task, but individuals with right parietal lobe dysfunction can have significant difficulties in this task of which they are completely unaware. For example, consider the story of Susannah Cahalan, author of the best-selling book, *Brain on Fire* (2012). She developed significant medical and neurological difficulties that no health providers could identify, despite having numerous medical evaluations and neuroimaging procedures over several months. However, like many individuals with right parietal lobe disorders, when completing the clock drawing test, she placed all of the clock's numbers and hands on the right side of the circle, leaving the left side of the circle/clock completely blank (similar to Fig. 3). In essence, she was not aware of the left side of space. This simple clock drawing task detected a neurological basis for her illness (eventually diagnosed as encephalitis), whereas advanced neuroimaging did not.

In addition to having difficulties with tests such as clock drawing, individuals with acute and severe right parietal lobe disorders often

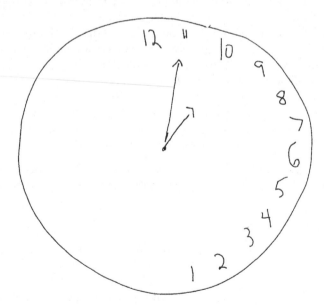

FIG. 3   Abnormal clock drawing.

demonstrate specific neglect syndromes in aspects of their daily lives that are difficult to comprehend. For example, some individuals with acute right parietal lobe stroke will eat only the food on the right side of their plate. Others will dress only the right side of their body. Eating food only on the right side of the plate may go relatively unnoticed by others (think of children who are picky eaters and move their food around on their plates). However, dressing only half of your body is likely to warrant some attention from family members and rehabilitation staff, particularly as you head out into the public where such lack of awareness is likely to lead to considerable commotion. Can you imagine dressing only the right side of your body and being unaware of doing so? Try to conceptualize how individuals with such disorders and symptoms must experience the world and to understand the sense of "self" that these individuals must have.

The previous examples demonstrate the relatively subtle manifestations of right parietal lobe dysfunction, although they can also have devastating implications (e.g., veering into oncoming traffic and dispensing the wrong dose of medication). However, right parietal lobe-related disorders of the self can produce even more unusual and bizarre symptoms.

7 WHOSE ARM IS THAT?

Again, consider whether or not you believe that your left arm is yours (i.e., that it is part of your "self"). It is difficult to conceive of a situation in which you would not do so. However, in fact, many individuals with right parietal lobe disorders will actually deny that the left side of the body (and usually the arm and less often the leg) is their own. This condition is known as asomatognosia and defined as the "ignorance of paralysis as a result of brain damage" (Merriam-Webster, 2017). How is it possible that individuals would be "ignorant" that all or part of the left side of their body was paralyzed?

The following excerpt is from a book by Feinberg and Keenan (2005, pp. 103–104) with a discussion between Dr. Feinberg and an elderly woman with a severe right hemisphere stroke and subsequent left hemiplegia, where she did not believe her left arm was actually hers:

> Feinberg: I want to ask you a question now. What is this over here (he pointed to her left arm)? Take a look at this over here. What is it?
> Patient: Your fingers.
> Feinberg: My fingers?
> Patient: Yes.
> Feinberg: Look at them again, take a good look. Ok … tell me what they are.
> Patient: Fingers … I see fingers and a pocket.
> Feinberg: Take a good look. What is it (tapping the back of her hand)?
> Patient: The back of your hand.

Feinberg: The back of my hand?
Patient: Yes.
Feinberg: Suppose I told you this was your hand?
Patient: I wouldn't believe you.
Feinberg: You wouldn't believe me?
Patient: No, no.
Feinberg: This is your hand.
Patient: No.
Feinberg: Look, here's your right hand and here's your left hand.
Patient: Ok.
Feinberg: Now, what's this (holding out her left hand)?
Patient: The back of your hand!

Consider the sense of "self" that this woman exhibits. What would your "self" be if you did not believe that your left arm, attached to your left shoulder, was a part of you? This is hard to conceive, but it illustrates that different regions of the brain construct a sense of self based on different neuropsychological processes. Change these self-oriented processes due to different neurological conditions, and it is clear that our "self" is not a concrete, unchangeable construct. Rather, our self is (or can be) an ever-changing concept that is dependent on neuropsychological processes that influence our perception of the world based on physical, psychological, and autobiographical constructs of self. Additionally, she insisted repeatedly that her left hand belonged to the doctor. She cognitively and rationally knew that it was a left hand, but due to her neurological condition, she believed that her left hand (and arm) was no longer part of her "self" and hence her belief that it must belong to the doctor's body.

8 KNOW THYSELF

These examples illustrate some of the ways that right parietal lobe-related dysfunctions reflect disorders of the physical self. That is, one can have a diminished ability to recognize the physical characteristics of their self, including being unaware of the position of their body in space or even that the left side of their body is even a part of their physical self. Other research with persons with right parietal lobe disorders indicates that there are also disorders of the "psychological self." The psychological self can be described as the sense of who we are in terms of personality characteristics, cognitive skills/thoughts, and behavioral traits.

To understand this psychological sense of self, consider the following. Remember a time when you had to introduce yourself to others (e.g., fellow freshmen at a college orientation, a person you're meeting on a blind date, and fellow participants at a mandated business retreat). How do you introduce yourself (i.e., your "self")? You will assume that you know your strengths and weaknesses better than anyone else. You may describe your best attributes as a talented musician, gifted athlete, accomplished writer, thoughtful conversationalist, empathetic friend, or even spiritual seeker. If asked to describe your weaknesses, you may admit that you are a terrible dancer, bumbling public speaker, or lacking in social graces. The point is that everyone develops a sense of who they are in terms of their personality characteristics and attributes (i.e., they know their "self"). It is generally accepted that people know what they are or are not good at (i.e., we have a sense of our positive and negative personal attributes). Some of us may overestimate our positive attributes (e.g., thinking that one is a great tennis player but actually being very average at best). Conversely, some of us may overestimate our negative attributes and weaknesses (e.g., believing that you do not draw very well but are much more talented than other artists). Taking away these relatively minor and normal variations in self-perception, most individuals generally have a good sense of the strengths and weaknesses associated with their psychological "self."

Now, consider if it is possible that our sense of who we are in terms of personality characteristics (i.e., our psychological self) can change based on neurological injury. Obviously, physical, circumstantial, and random events can bring about immediate changes in our sense of psychological self (e.g., going from wealthy and successful business person to financially destitute pauper). However, is it possible that a neurological injury can also dramatically change our sense of psychological self? Suppose for a minute that one of your best friends lost the ability to objectively describe their own strengths and weaknesses. What if your most uncoordinated friend suddenly believed they were the most talented player in the noon basketball game at the local YMCA? What if your socially awkward college roommate suddenly developed the confidence of a successful movie star and attempted to become the master of ceremonies at a sorority social event? What if your delightful but intellectually limited partner suddenly quit work as a physical laborer and decided to become president of their own start-up computer software company? You may think they all lost their minds. The point is people generally know their psychological "self" and their capabilities.

However, case studies of individuals with specific dysfunctions of the right hemisphere indicate that such individuals apparently have lost the ability to "know themselves," at least compared with what their families and friends think. This inability to identify one's personal strengths and weaknesses is known as anosognosia, which is defined as "an inability or refusal to recognize a defect or disorder that is clinically evident" (Merriam-Webster, 2017).

It is nearly impossible to list first-person accounts of anosognosia given that people with the disorder are not aware that they have the disorder (i.e., they are unaware that they are unaware of their deficit). Anosognosia is an unusual disorder compared with other neuropsychological disorders, as individuals with other cognitive disorders are aware of their neuropsychological weaknesses and can comment on them directly. For example, individuals with acquired dyslexia (i.e., inability to read, associated with the dysfunction of specific left hemisphere regions) can tell you that they can no longer read. Similarly, individuals with acquired attention disorders associated with specific frontal lobe injuries can tell you that they are unable to maintain attention and are easily distracted. Individuals with disorders of the temporal lobes and limbic system can have significant memory problems, of which they can easily identify and report. However, individuals with an inability to accurately recognize their own attributes are not able to process their strengths and weaknesses (at least not as others who knew them prior to the onset of their disorder now see them). Just as you firmly believe that you know who "you" (i.e., yourself) are as you read this chapter, individuals with anosognosia also firmly believe that they also know who "they" (i.e., their self) are. This raises an interesting question from both neurological and empirical perspectives: who has a better understanding of a person's self? The individuals themselves or family members and friends who are aware that an individual's characteristics have significantly changed? It is an interesting question with uncertain answers.

Individuals can present with varying degrees of anosognosia. In most clinical settings, it is generally accepted that it is necessary to take the patient at their word (i.e., why not believe a person referred for an evaluation of brain functioning when they describe their situation?). However, interviewing individuals with anosognosia can be problematic without the input of family members and friends and particularly when you only have brief interaction with the affected individual.

Consider the following case of an individual evaluated at the neuropsychology clinic at the University of Missouri. "William" (not his real name) was a successful accountant living in a small, rural area of Missouri. He had earned a master's degree in accounting and had worked as a partner in a successful accounting firm for years. In addition to his professional success, he was happily married and an active member in his community. Unfortunately, William was involved in a serious car accident and suffered a severe brain injury that primarily injured his right cerebral hemisphere and particularly the right parietal lobe (i.e., multiple hemorrhages/contusions). He was rendered unconscious for several hours following the accident and did not remember the day before the accident or several weeks following, and he had to undergo extensive rehabilitation over the next year.

Despite being very bright and earning a master's degree prior to his traumatic brain injury (TBI), William experienced significant

II. THE "SELF" AND SELFLESSNESS

neuropsychological weaknesses following his injury. Although William was estimated to have had above average neuropsychological abilities prior to his accident (i.e., stronger than 75% of his same age peers), 2 years after the accident, his intelligence and neuropsychological abilities had generally fallen to the borderline range (i.e., stronger than only 3% of his same age peers). In addition, as is common with severe cases of TBI, William was unable to return to work as an accountant and as a result was awarded social security disability income. In addition, he and his wife divorced subsequent to his brain injury, primarily because of the significant emotional, behavioral, and personality changes William displayed after his accident, particularly because of his apparent inability or unwillingness, as perceived by his wife, to recognize how he had changed. Furthermore, it became necessary for him to live in a residential facility for persons with severe TBI, given his inability to independently care for himself after his accident.

Despite the serious nature of his TBI, the resulting neuropsychological impairments, and significant lifestyle changes, William continued to insist that he had experienced no changes following his accident and was fully capable of returning to his previous employment. He also believed that he should be able to live independently and that he would be able to convince his wife to reunite with her if he was given the chance. Just as you or I would swear that we know our "selves," including our personality characteristics, cognitive abilities, and behavioral traits, William was 100% confident that he too knew who "he" was, including his ability to continue doing all the things he had done prior to his TBI. Obviously, William's sense of his "self" was significantly changed following his severe TBI, as is the case for many individuals with severe injury to the right parietal lobe.

9 MIRROR, MIRROR, …

One of the more interesting disorders of the self is known as mirror misidentification syndrome. To fully understand this disorder, consider the following. Look in a mirror. Do you recognize yourself? Can you ever conceive of a situation in which you would not recognize yourself in a mirror? Actually, it appears that there is a neuropsychological process associated with recognizing and remembering your own face. But what would happen if you experienced injury to the part of the brain associated with the neuropsychological process of recognizing yourself?

Breen et al. (2000, pp. 83–84) published an article that presented the interesting case of an 87-year-old man who suffered a right hemisphere stroke and subsequent mirror misidentification disorder. Neuropsychological testing indicated that his left hemisphere abilities were generally intact but that he had significant right hemisphere impairments. This individual could no longer recognize himself in the mirror, and he treated his reflection as if it was someone who followed him around everywhere (e.g., in his home, in the car, and at the shopping center). When he attempted to communicate with his reflection, he could not understand why his reflection did not talk back to him. It did not appear to cause him distress, other than the one time he saw his own reflection in the full-length bedroom mirror when he was in bed with his wife. He could not understand that this was actually his own reflection despite numerous conversations about this incident with his family. Following is an excerpt of a conversation between a researcher and this individual related to his experiences as he looked at his own reflection in a mirror:

Examiner: Who is that?
Patient: That's not me. It hits me straight away. First of all, I didn't like his face at all, but I've got used to his face, and I'll have a smile with him if I'm in the bathroom for a wash or something, but it's not me.
Examiner: It's not you?
Patient: No.
Examiner: So this is the person you see in your house?
Patient: Yes.
Examiner: And it's not you.
Patient: No.
Examiner: What does this person look like?
Patient: Well, he very much looks like me. I guess he could pass for (me).
Examiner: He does look like you?
Patient: Yes, I see that. He's not a bad-looking fellow.
Examiner: So, what does he look like? Can you describe him?
Patient: No better than that (indicating toward his reflection in the mirror).
Examiner: Does he wear glasses?
Patient: I think he does, I think he does. Yes, he does wear glasses (the patient did wear glasses).
Examiner: What color is his hair?
Patient: I don't think he is as white as I am (patient had white hair and was balding).
Examiner: Is he going a bit bald, or does he have a full head of hair?
Patient: Ok, I think he's about the same as mine as far as hair coloring is concerned.

The examiner (named Nora) then describes a situation in which she stands next to the patient in front of a mirror, with the reflections of both herself and patient visible to the patient (Breen et al., 2000, pp. 84–85).

Examiner (pointing to her own reflection): Who is this, next to the person?
Patient: I don't know.

II. THE "SELF" AND SELFLESSNESS

Examiner: Who does it look like? Have you seen this person in here before?
Patient: That's you.
Examiner: That's me?
Patient: Yes.
Examiner: Me, here? What's my name?
Patient: I don't know, oh yes, it's Nora.
Examiner: Nora, that's right? So that's me in the mirror?
Patient: Yes.
Examiner: That's my reflection?
Patient: Yes.
Examiner: And who is that (pointing to the patient's reflection)?
Patient: I don't know what you would call him. It makes me a bit sick because he moves about freely with us. I don't be too friendly because I don't see it does him any good.

Feinberg and Keenan (2005) describe a similar case in which a woman with significant periventricular white matter changes in the bilateral parieto-occipital watershed regions, particularly involving the right posterior parietal region. This woman began to have hallucinations that a "little girl" was living in her home, whom she saw in mirrors and other reflective surfaces. When she was provided with a mirror to look into, she became animated, stating "That girl! What is she doing here? I thought she was only back home!" When asked to describe the girl, she stated that the girl resembled her but was much younger than herself (i.e., that the girl in the mirror looked like herself when she was younger). When it was pointed out that she was looking at a reflection of herself in a mirror, the patient adamantly denied it, stating "It's the girl!" The examiner then tilted the mirror so that the patient could see the examiner and asked who that was. The patient was "easily able to recognize the examiner but unable to recognize herself" (p. 132).

These cases of mirror misidentification syndrome show that individuals with primarily right parietal lobe dysfunction can have difficulties recognizing themselves, but not other people, in the mirror. This suggests that their difficulties are not associated with an inability to recognize any face (a condition known as prosopagnosia), but are specific to the inability to recognize only their own face (i.e., self-facial recognition).

10 IMPOSTERS

There are many other "disorders of the self" that are extremely interesting to study but equally difficult to fully understand, many of which are characterized as psychiatric disorders. One such category of psychiatric conditions includes delusional misidentification syndromes

(DMSs), which involve a distorted sense of "personal relatedness" of the self to others. In general, consistent with other disorders of the self, these misidentification disorders have been shown to be primarily related to the dysfunction of the right hemisphere and right parietal lobe (Feinberg, DeLuca, Giacino, Roane, and Solms, 2005). For example, Capgras syndrome, first characterized in 1923, is described as a disorder in which the individual has a delusional belief that a person or persons known to them have been replaced by impostors. It has been described as involving "neurological perturbations of the self in which brain dysfunction creates a transformation of *personal significance*" (Feinberg et al., 2005, p. 101). Stated simply, individuals have changes in the experience of "personal relatedness" between themselves and other persons, objects, events, or experiences. For example, Alexander et al. (1979) reported the case of a 44-year-old man with TBI-related right frontotemporal encephalomalacia (i.e., the softening or loss of brain tissue after cerebral obstruction or limitation of blood flow, infection, blunt trauma, or other injuries) who claimed that his wife and five children had been replaced by impostors. He could recognize the physical features of his family, but he did relate to them as "my" family. Another case described by Stanton et al. (1982) reports a 31-year-old man with a right frontotemporal and parietal brain injury who claimed that his parents, siblings, and friends were not real but were actually substitutes. Can you imagine that you could ever believe that your family had been replaced by impostors?

There are other examples of other highly unusual DMSs that are primarily related to right hemisphere dysfunction, including Fregoli's and Cotard's syndromes. Fregoli's syndrome involves an overpersonalization of relationships in which the affected person believes that individuals who have no relationship to them are in fact well known to them. Ruff and Volpe (1981) described a case in which a 60-year-old woman who had evacuation of a right frontal subdural hematoma (i.e., a swelling of clotted blood) stated that she believed that the person in her hospital room in the bed next to hers was her husband. In another case, a 41-year-old man was hospitalized for right frontal intracranial and subarachnoid hemorrhages. Although he was in a rehabilitation hospital recovering from his surgery, he claimed that he was at work and that the hospital staff was individuals from his workplace (Feinberg & Keenan, 2005). These individuals all had disorders in which they overpersonalized the perceived relationship of their "self" to others.

A syndrome that is even more difficult to understand is Cotard's syndrome in which individuals maintain the delusional belief that they are dead, do not exist, are putrefying, or have lost their organs or blood. Young and Leafhead (1996) described the case of an individual who incurred a primarily right hemisphere brain injury in a motorcycle accident.

II. THE "SELF" AND SELFLESSNESS

This individual reported general feelings of unreality and of being dead! After he was discharged from a hospital in Scotland, he went to South Africa with his mother. He believed that he had been taken to hell, which he related to the heat he experienced there and that he had died of septicemia (i.e., infection of the blood), AIDS, or from an overdose of a yellow fever injection. He believed that he had "borrowed [his] mother's spirit to show [him] around hell" and that during such time she was asleep in Scotland.

Although complex, DMSs are believed to be primarily related to right hemisphere impairments. In a review of 29 cases of DMS, it was reported that almost half of the cases (48%) had exclusive right hemisphere damage, while no cases had exclusively left hemisphere damage (Feinberg et al., 2005). Furthermore, it was noted that 45% of these individuals had right parietal lobe damage and only 3% had left parietal lobe damage. It was concluded that the "data indicate an overwhelming bias toward right hemisphere damage in terms of DMS" (p. 108). A possible explanation for the manner by which these misidentification disorders manifest was offered, suggesting that right hemisphere damage is associated with disturbances of self/ego boundaries, which can negatively affect both physical self-other boundaries and representations of psychological ego boundaries.

11 WHOSE THOUGHT IS THAT?

Although the previously described disorders of the self are primarily related to neurological injuries and diseases (e.g., traumatic brain injuries, seizures, strokes, and brain tumors), there are also several psychiatric conditions in which individuals experience a disordered sense of the self. For example, schizophrenia is a psychiatric condition that involves impairments in thinking and perception, primarily experienced in terms of hallucinations or delusions. Hallucinations are characterized by thoughts and perceptions that are not real (although they are "real" to the person with schizophrenia), such as hearing voices, seeing images, or feeling sensations that others cannot. Delusions, on the other hand, involve the creation of complex belief systems that are not based in reality (e.g., that they possess unique religious powers) and often involve feelings of paranoia (e.g., that others are conspiring against them and that the government is monitoring their activities through electronic devices).

Hallucinations are most relevant to our model, however, as individuals with schizophrenia generate thoughts, but do not experience them as originating from within their "self." That is, they experience a thought, but not as "my" thought. Given that they are unaware that it is their "self" that is generating the thought, they apparently misperceive the thought as belonging to other people (familiar or unfamiliar) or objects (e.g., television, radio, implanted electronic devices, divine beings, and aliens). Consider the experiences of John Nash, the Princeton professor, mathematician, Nobel Prize winner, and the subject of the 2001 movie, *A Beautiful Mind*, who was diagnosed with schizophrenia. In describing his auditory hallucinations, he stated that he thought of the voices he heard as originating from aliens or possibly angels but stated that during periods of good health, he came to understand his delusions as originating from his subconscious.

To date, it has been difficult to understand the specific neuropsychological mechanisms associated with schizophrenia as it is associated with many different symptoms, neural networks, cognitive processes, and neurotransmitters. It has been conceptualized primarily as a chemical disorder, which is why it is usually treated with medications. However, schizophrenia is increasingly being regarded and researched as a disorder of self-awareness (Beitman & Nair, 2004; Robinson, Wagner, and Northoff, 2016; Sass, 1998, 2001, 2014; Sass & Parnass, 2003), and several studies have focused on abnormalities within right hemisphere structures and functions (Barnett, Kirk, and Corballis, 2005; Mitchell & Crow, 2005).

Robinson et al. (2016) indicate that "schizophrenia is a disturbance of the self" (p. 270), and Sass (1998) describes the self, as it relates to schizophrenia, as an "*unseen point of origin for experience*, thought, and action, as a medium of awareness, source of activity, or general directedness toward the world" (p. 6, emphasis added). Individuals suffering from schizophrenia often lose this point of perception (i.e., self), whereby their thoughts are not integrated into the experience of a self. As a result, thoughts are perceived as coming from others. Just as individuals with asomatognosia cannot recognize their arm as their own (i.e., not "my" arm), individuals with schizophrenia cannot recognize a thought as their own (i.e., not "my" thought).

The conceptualization of schizophrenia as a disorder of the self is supported by research that has identified childhood predictors of adult schizophrenia. Recent theories are suggesting that disturbances in the manner by which the self is processed in children and adolescents may predict the development of schizophrenia in adults (Brent, Seidman, Thermenos, Holt, and Keshavan, 2014). It has been suggested that early abnormal maturational processes in the frontal, temporal, and parietal lobes that are involved in self-processing may lead to subtle disruptions of "self" for persons at risk for schizophrenia. Specifically, it has been proposed that schizophrenia may be related to difficulties in the ability to develop an "integrated sense of self" (Brent et al., 2014, p. 77). In fact, impaired "self-reflection" (described

II. THE "SELF" AND SELFLESSNESS

as an altered sense of self) is among the earliest reported symptoms in individuals who later develop schizophrenia (Klosterkotter, Hellmich, Steinmeyer, & Schultze-Lutter, 2001; Poulton, Caspi, Moffitt, Cannon, Murray, & Harrington, 2000). These theories are consistent with studies that suggest that the sense of self is a neurodevelopmental process that develops in children over time and starting between the ages of 18 and 24 months (Lewis, Sullivan, Stanger, & Weiss, 1989; also see Chapter 4), suggesting that early weaknesses in the ability to develop a sense of self may be the root cause of psychiatric disorders such as schizophrenia.

12 THE SELF AND OTHER PSYCHIATRIC CONDITIONS

In addition to schizophrenia, other psychiatric conditions are conceptualized as disorders of the self, including depersonalization disorder and dissociative identify disorder. Depersonalization disorder is a condition in which individuals report feeling "as if" their experiences are not real (Radovic & Radovic, 2002). For example, they may feel *as if* they are an outside observer of their own thoughts, feelings, and body or *as if* they are a robot and not in control of their thoughts, emotions, and movements. Sometimes, they feel *as if* their sense of bodily limbs is distorted (i.e., their arms and legs are shortened or lengthened) or that their memories are not their own. The experience of depersonalization disorder has been described by one individual with the disorder as follows:

> Depersonalization is terrifying because it appears to completely erase your identity, *your sense of self*, and as a result your values in life. (*Hanson, 2018, emphasis added*)

Relatedly, dissociative identity disorder (previously known as multiple personality disorder) is another psychiatric condition that involves complex symptoms and has multiple causes (that are primarily related to extensive childhood psychological or sexual abuse). The *Diagnostic and Statistical Manual-5* (the professional source book for psychiatric diagnoses) indicates that dissociative identity disorder involves a "disruption in identity (that) involves marked discontinuity in *sense of self* and agency" (American Psychiatric Association, 2013, p. 292, emphasis added). In simple terms, individuals develop multiple personalities (i.e., multiple selves) that are distinct, different, and often unaware of one another. Although controversial, this interesting disorder has been the subject of several well-known books and movies including *The Three Faces of Eve* (based on the book by Thigpen & Cleckley, 1957) and *Sybil: The Classic True Story of a Woman Possessed by Sixteen Personalities* (Schreiber, 1973). To date, an estimated 40,000 cases of dissociative identity disorder have been diagnosed (Lynn, et al., 2012; Maldonado & Spiegel, 2008), although there is ongoing controversy as to the existence of the disorder and its multiple influences (e.g., trauma, genetics, culture, and therapist-induced).

Dissociative identity disorder is fairly easy to understand as a disorder of the self. Although rare, individuals with this disorder have a disintegration of the core personality (i.e., self) into multiple personalities (i.e., separate selves) as a way to cope with significant psychological stressors. Thus, compared with previously described disorders of the self, the sense of self of individuals with dissociative identity disorder is extended into multiple selves, rather than reflecting a decreased sense of self. Regardless of these complexities and the lack of clarity regarding the specificity of diagnoses, it is clear that individuals can experience psychiatric disorders that involve a distorted sense of self where multiple selves are generated.

13 DISORDERED SELF (OR SELFLESSNESS)?

The case studies described in this chapter have focused on the identification of neurological "disorders" of the self, such as left-sided neglect, asomatognosia, anosognosia, delusional misidentification disorders, and other psychiatric conditions. However, it appears that such conditions associated specifically with weaknesses in self-orientation do not always operate as "disorders." To work toward exploring what happens when a reduced focus on the self occurs without neurological injury, Chapter 4 reviews current empirical research on the manner by which the brain neuropsychologically processes a sense of the self.

4

Neuroscience of the Self

The self isn't something outside of experience, hidden in the brain or in some immaterial realm. It is an experiential process that is subject to constant change. We enact a self in the process of awareness, and this self comes and goes depending on how we are aware. (**Thompson, 2015, p. xxxi**)

The study of the self was largely in the domain of philosophers (e.g., Descartes and Kant) until the beginning of the 20th century. Up until that time, the self was generally conceived of as a permanent entity, the essence of a person, whether seen as the mind of the thinker or as the transcendent ego. In his seminal work the *Principles of Psychology*, William James questioned

© 2019 Elsevier Inc. All rights reserved.

these commonly accepted philosophical notions of the self. In his abridged version of this famous text, *Psychology: Briefer Course* (James, 1984/1892), James devoted a revised chapter to "The Self." He described the self as involving "passing states of consciousness" that are uniquely our own and found it highly suspect that traditional ideas viewed the self as something that is specific and unchanging. "The logical conclusion seems to be that the states of consciousness are all that psychology needs to do her work with. Metaphysics or theology may prove the soul to exist, but for psychology the hypothesis of such a substantial principle of unity is superfluous" (James, 1984, p. 181). However, James' theories were only speculative and could not be empirically validated until functional neuroimaging techniques (e.g., PET, fMRI, and SPECT) were developed in the last quarter of the 20th century. These technologies allowed neuroscientists to study the neurophysiology of specific neuropsychological processes in real time. For the first time, scientists could see what happens in the human brain when we think of our "self," with results suggesting that self is more of a neuropsychological process (i.e., a sense of self) than a permanent, irreducible entity (i.e., the self).

1 THE NEUROPSYCHOLOGICAL SELF

As is the case for most new disciplines, the neuroscientific study of the self has been somewhat disjointed. Without a clear understanding of what the "self" is, it has been difficult to study. However, there is "widespread agreement in neuroscience" that the body has a "central role in giving rise to the self" (Ananthaswamy, 2015, p. 140). Although the edited volume, *The Lost Self: Pathologies of the Brain* (Feinberg & Keenan, 2005), succinctly summarized different neurological disorders of the self, an equivalent book on the neuropsychological processing of the self has yet to be written. The most concise collection of empirical studies of the self was published in a special issue of *Consciousness and Cognition* (2011) and included diverse topics such as nested neural hierarchies of the self, brain imaging of the self, self/other differentiation, embodiment of the self, cerebral disconnection and the self, visual self-facial recognition, depersonalization and the self, phenotypes of the self, and "mindblindness" and the self in autism. The diverse nature of the articles illustrates the difficulties inherent in defining, measuring, and researching the self, whether as a neuropsychological process or a permanent entity.

From a neuropsychological perspective, we argue that the self is better understood as a process rather than a thing. To explore this orientation, it is helpful to review the experimental research that has been conducted regarding the nature of the self. The previous chapter has explored how regions of the brain, primarily within the right hemisphere, are associated with creating a sense of self, and that injury to those regions lead to conditions in which the sense of self is dramatically reduced. Conditions such as asomatognosia, anosognosia, and mirror misidentification disorder involve a diminished sense of self where individuals deny that certain body parts and physical/personality characteristics are their own. However, these "disorders of the self" raise other interesting questions. For example, if brain processes can be altered so that the typical experience of the self can be reduced and denied (e.g., not "my" arm, not "my" reflection, and not "my" personality characteristic), is it possible for the experience of the self to be distorted in other ways? Several research studies provide important clues into these questions.

2 IS THAT YOUR (RUBBER) HAND?

A series of studies over the past 20 years suggests that the brain can be convinced into believing that a rubber hand is part of one's body. The "rubber hand" experiment was first published in 1998 (Botvinick & Cohen, 1998) with numerous replications completed since then (just do an Internet search for "rubber hand experiment"). In these studies individuals sit with both of their hands extended on a table in front of them. However, the left hand is extended a bit to the left and then hidden behind a screen so that it cannot be seen. A rubber hand is then placed in front of the body, in a position that is symmetrical to the placement of their right hand (see Fig. 1). The participants are asked to fix their eyes on the rubber hand and the examiner then uses brushes to simultaneously stroke the participant's left hand and the rubber hand. It is likely that most of us would believe that the participants should be able to distinguish between their own hand and the rubber hand. However, research has consistently shown that individuals actually report the sensation that the rubber hand is theirs (i.e., that it is part of

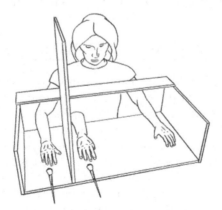

FIG. 1   Rubber hand experiment. *From de Haan, A. M., Van Stralen, H. E., Smit, M., Keizer, A., Van der Stigchel, S., & Dijkerman, H. C. (2017). No consistent cooling of the real hand in the rubber hand illusion. Acta Psychologica, 179, 68–77. Copyright Elsevier.*

their "self"). The title of Botvinick and Cohen's study describes the nature of the phenomenon perfectly: "Rubber hands 'feel' touch that eyes see." What can explain this apparent misperception? The researchers suggest that "the body is distinguished from other objects as belonging to the self by its participation in specific forms of *intermodal perceptual correlation*" (Botvinick & Cohen, 1998, p. 756, emphasis added). In simple terms, the authors are indicating that the brain engages a process by which the different sensory perceptions are integrated into a sense of the self. The brain perceives "intermodal" sensory perceptions (e.g., the feeling of the left hand being touched, the sight of the rubber hand being stroked, and the proprioceptive sense of the location of the left hand relative to the body) causing it to rearticulate the sense of bodily self. Notably, the ability to perceive the different senses is intact, but the neuropsychological process by which this sensory information is integrated in to a sense of self has been altered. Although the brain feels that the left hand is being touched in a specific spatial location relative to the rest of the body, it sees the rubber hand (but not the actual hand) being touched. When the brain attempts to integrate these sensory experiences to create a unified sense of self (i.e., I see and feel my hand being touched in a specific spatial location), it perceives that rubber hand as part of its self. That is, the overall sense of self has been altered to include nonbody parts because of conflicting sensory information being input simultaneously.

Comparatively, it is important to note here that individuals with clinical disorders of the self such as asomatognosia, anosognosia, or mirror misidentification disorder typically have extensive right hemisphere brain damage. However, those individuals who perceive that a rubber hand is part of their self are cognitively intact. It appears that one does not have to have brain damage to experience an altered sense of self.

3 IS THAT MY (AMPUTATED) ARM?

Other fascinating research also indicates that the sense of self can be altered based on conflicting sensory information that is presented to the brain. In a study of 10 individuals, Ramachandran and Rogers-Ramachandran (1996) demonstrated that individuals with amputated arms could be convinced into thinking that they could feel sensations of touch and see movement in their amputated limbs. In their study, individuals with amputated arms were asked to put their intact arm in a specially designed box with mirrors. The mirrors were arranged so that the intact arm was reflected in the mirror, which gave the impression that the amputated arm was also intact (see Fig. 2). Their intact arm was then touched while they continued looking into the mirror,

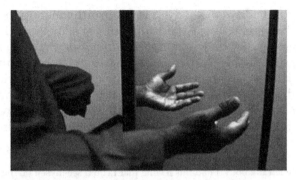

FIG. 2   Amputated limb mirror experiment. *From Ramachandran, V. S., & Altschuler, E. L. (2009). The use of visual feedback, in particular mirror visual feedback, in restoring brain function. Brain, 132 (7), 1693–1710.*

and several of these individuals reported that they felt as if it was their amputated limb that was actually being touched. Based on this conflicting sensory information, their brains perceived that it was their amputated arm that was being touched, thus creating a temporarily distorted sense of the self. Similar to the rubber hand experiment, it appears that the ability to integrate sensory information into a sense of self can be altered.

These studies suggest that the brain perceives various sensory sensations (i.e., visual, tactile, and proprioceptive) but that the ability to integrate them into a sense of self can be altered by the presentation of conflicting, ambiguous, or confusing sensory information. In these experiments, the brain perceived the experience of touch to a hand, which was in conflict with visual information (i.e., a rubber hand or a reflection in the mirror). The neuropsychological processes that integrate these visual, tactile, and proprioceptive sensory stimuli gave the impression that the rubber hand, or the reflected arm in the mirror, was therefore a part of the self. This type of research suggests that conscious awareness of the self is modulated and remodulated on a continual basis neuropsychologically and that the sense of self operates as an ongoing process that is neither static nor permanently fixed.

4 OUT-OF-BODY EXPERIENCES

There are other intriguing studies that also suggest that the "sense of self" is a neuropsychological process that can be altered, with one of the most interesting conditions involving out-of-body experiences (OBEs). These experiences typically involve a sensation of floating outside one's body or the feeling of perceiving one's physical body as if from a place outside of the body. It is noteworthy that up to 10% of the population has reported experiencing OBEs at some time during their lives, suggesting that they may not be unique or unusual experiences (Blackmore, 1982). These experiences vary but are commonly reported by individuals undergoing surgery and particularly individuals who have near-death experiences. For example, the actress Jane Seymour (known for her TV role as Dr. Quinn Medicine Woman) reported an OBE related to a serious illness she experienced in 1988, which she described as follows:

> I literally left my body. I had this feeling that I could see myself on the bed, with people grouped around me. I remember them all trying to resuscitate me. I was above them, in the corner of the room looking down. I saw people putting needles in me, trying to hold me down, doing things. (*Seymour, 2015*)

Another story illustrates how individuals can have OBEs without illness or injury but instead at times of high stress. During her first night on duty, a police officer was involved in pursuit of an individual with a weapon. She described the following experience:

> When I and three other officers stopped the vehicle and started getting [to] the suspect ... I was afraid. I promptly went out of my body and up into the air maybe 20 feet above the scene. I remained there, extremely calm, while I watched the entire procedure – including watching myself do exactly what I had been trained to do. (*Alvarado, 2000, p. 183*)

Others have reported OBEs as occurring while they slept, during meditative practices, during high-altitude climbs, or while flying planes. Numerous anecdotal stories and research on pilots suggest that the significant gravitational forces applied to the body and brain during high-speed, high-altitude flights can lead to OBEs (Whinnery & Whinnery, 1990). Specifically, pilots have frequently reported feeling as if they are on the wing of their airplane or above and behind their self in the cockpit, watching their self flying the plane.

The origins of OBEs have been explained in several different ways, including theories suggesting that they are related to supernatural events, hypoxic events, or psychiatric conditions involving sensory-related hallucinations. However, regardless of the purported cause of OBEs, it is noteworthy that most studies indicate that OBEs involve significant alterations in ego and temporal–spatial relationships (i.e., a different sense of "self"; Greyson, 2003). In fact, OBEs are often described in terms of the self but a self that is experienced in an altered sense of space. Specifically, one's sense of a self exists (i.e., the feeling of "me"), but the spatial positioning of one's self-awareness has been altered (e.g., "I" am above "my" physical body).

Possibly the most well-known OBE research has been conducted by Otto Blanke and his colleagues in Switzerland. In a descriptive study,

Blanke, Landis, Spinelli, and Seeck (2004) studied six individuals with neurological disorders (primarily seizures) who reported experiencing OBEs. The OBE of one of their subjects was reported as follows:

> Patient 1 ... saw herself (entire body as lying on the ground, facing up) and some unknown people (some were standing around her body, others were moving around) below. Initially, she felt as being 'above her real body', but that she was rapidly rising higher. She felt as if her elevated body was in the horizontal position, but did not see any part of it. The visual scene always took place outdoors and was described as 'a green meadow or hill'. *(Blanke et al., 2004, p. 245)*

Neurological evaluations and interviews indicated that the OBEs were primarily associated with lesions in the temporoparietal junction for five of the six participants. Similar to conclusions drawn to explain the rubber hand experiments, these authors speculated that OBEs were related to a "failure to integrate proprioception, tactile, and visual information with respect to one's own body" (p. 243). It appeared that these individuals were accurately processing different sensory experiences, but abnormalities in other neuropsychological processes were leading to a failure to integrate these sensations into a "normal" sense of self. In these cases, due to distorted neuropsychological processes based on damage to the temporoparietal junction, the typical "sense of self" was altered so that it was experienced from a different spatial perspective. That is, these individuals reported a sense of self but one that was located in a different spatial position than that of the bodily self (i.e., an OBE).

5 INDUCING OBES

In addition to reporting anecdotal cases of spontaneous OBEs, several researchers have conducted experiments to determine if OBEs (or similar experiences) can be induced in neurologically intact individuals. Their studies were based on the premise that if such experiences are neurologically based, then they should be able to be artificially induced. Several OBE studies have suggested that the temporoparietal junction, particularly in the right hemisphere, is associated with such experiences.

For example, in one study, Blanke and colleagues applied transcranial magnetic stimulation (TMS) to the right temporoparietal junction of seven healthy individuals to distort their ability to orient themselves in space (Blanke et al., 2005). As hypothesized, each of these individuals demonstrated an impaired ability to mentally transform the position of their own bodies in space, although no such effects were found in their ability to mentally orient other objects that were shown to them.

In a study similar to the rubber hand experiment, Blanke and colleagues used virtual reality techniques to disrupt the spatial unity between the self and the body (Lenggenhager, Tadi, Metzinger, & Blanke, 2007). Rather than having participants look at a rubber hand, in this experiment, two video cameras were placed approximately two meters behind the participants, and these camera images were presented in a pair of head-mounted displays so as to create the sensation that the participants were watching themselves from behind. In the experiment, the participants' actual and virtual bodies (i.e., their backs) were then stroked in a synchronous fashion (see Fig. 3). After this occurred, the participants were asked to identify where their bodies had been located during the experiment. Based on conflicting visual and somatosensory perceptions, the participants inaccurately localized themselves toward the virtual body to a position outside their actual bodily borders.

At the same time that Blanke and his colleagues were conducting OBE research in Switzerland, researchers in Sweden also induced OBEs in healthy subjects (Ehrsson, 2007) and obtained similar results. Similar to the Swiss studies, participants wore a pair of head-mounted video displays, but in this study, the researchers stroked the chest of the participants and the virtual bodies. Once again, the research participants reported the experience of sitting behind their physical bodies and looking at them from that location. In all these OBE studies, researchers concluded that bodily self-consciousness can be studied experimentally and is based on the multisensory and cognitive processing of bodily information.

Notably, there is a wide range of other OBE and disembodiment syndromes that have been recognized. For example, a condition known as autoscopy involves seeing one's body in an extrapersonal space

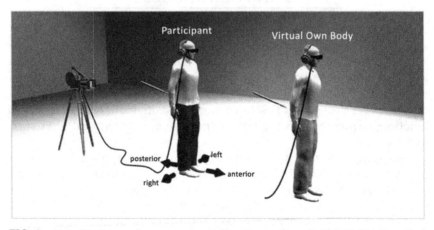

FIG. 3 Out-of-body experience experiment. *From Lenggenhager, B., Tadi, T., Metzinger, T., & Blanke, O. (2007). Video ergo sum: Manipulating bodily self-consciousness. Science, 317, 1096–1099.*

outside of one's immediate physical body, during which the person experiences a close affinity with this external (autoscopic) body (see Blanke (2012) for a review of various multisensory self-identification and self-location disorders and anomalies). Another disorder of body ownership occurs in a condition called somatoparaphrenia in which one believes that a limb or limbs do not belong to oneself. Importantly, both of these disorders are directly associated with damage to the right parietal lobe of the brain (Giummarra, Gibson, Georgiou-Karistianis, & Bradshaw, 2008). The wider point here is that there are a considerable range of neurological conditions that can cause anomalies in self-identification and self-location, most of which involve the right hemisphere and parietal lobe.

6 INTEGRATING SENSORY EXPERIENCES INTO A SENSE OF SELF

So what do these scientific experiments tell us about the neuropsychological process of creating a sense of self? What is the "self" if our perceptual abilities can be altered so that we believe that rubber hands or amputated limbs are part of our "self" or if the bodily sense of self can be manipulated through conflicting sensory inputs? It is commonly understood that the brain perceives different sensory information in terms of sight, sound, taste, smell, touch, and spatial positioning. In addition to perceiving these different sensations, the brain also integrates these sensory experiences into an overall sense of self, as part of a unified experience that belongs to "me." It is commonly accepted that there is an experience of "me," which is the recipient of these sensations, and that this sense of "I/me/mine" has some form of coherence or individuality over time and through different circumstances. However, as the rubber hand and the OBE experiments illustrate, the ability of the brain to create a sense of self can be altered. It is important to note that the sensory perceptual abilities (i.e., visual, tactile, and proprioceptive sensations) of the individuals in the rubber hand and OBE studies were all intact, but their *ability to integrate these sensations* was distorted. This raises the following question: if the ability to perceive sensations is intact, what is the neuropsychological process that is impaired for these individuals? These studies suggest that it is a specific neuropsychological process that integrates these sensations into a coherent unified experience (i.e., a sense of self) and that these neuropsychological processes reside primarily in the right hemisphere.

Importantly, in order to understand how the brain creates a sense of self as a continually negotiated neuropsychological process, it may be helpful to consider how other neuropsychological processes can be similarly altered. For example, memory is a neuropsychological process that can be diminished as in the case of amnesia, in which individuals cannot recall past memories. We all experience various degrees of amnesia/forgetfulness at times, which involve a decreased ability to recall information. Neurological disorders such as Korsakoff's syndrome (a dementing condition associated with excessive alcohol consumption) are known to distort individual's memories but in a different manner than the traditional amnesias. Specifically, one of the hallmarks of Korsakoff's syndrome is confabulation, which is the tendency to recall information that is factually inaccurate. In common parlance, individuals with Korsakoff's syndrome appear to make up inaccurate stories about their past, which they believe are accurate. These individuals are not lying, they are confabulating. They recall memories, but their process of remembering is distorted and works incorrectly. These types of conditions show how the neuropsychological processes of memorization can be both *diminished* (e.g., through amnesia) or *distorted* (e.g., through confabulation). Similarly, it appears that the neuropsychological processes that integrate sensory and mental experiences to create a sense of self can also be diminished (e.g., as seen in asomatognosia) or distorted (e.g., as per the rubber hand illusion).

7 PHYSICAL AND PSYCHOLOGICAL SELVES

Whereas the previous studies indicate that the right hemisphere integrates various sensory experiences into a general sense of self, other research has distinguished between distinct physical and psychological senses of the self. Studies have specifically indicated that the right hemisphere integrates various sensory experiences into an integrated sense of a physical self (e.g., my body, my arm, my face, and my voice) and various *mental experiences* into an integrated sense of a psychological self (e.g., my personality and my mental skills). Differences between the physical and psychological selves are best illustrated when considering disorders such as asomatognosia (i.e., a disorder of the physical self) and anosognosia (i.e., a disorder of the psychological self).

8 RIGHT HEMISPHERE AND THE PHYSICAL SELF

The majority of empirical studies on the neurological correlates of the physical self has focused on the manner by which the brain recognizes faces. This line of research developed, in part, because of anecdotal case studies of individuals with mirror misidentification disorder, namely, the inability to recognize one's own face in the mirror. Although methods of facial recognition in these studies vary considerably, most have focused on the manner by which specific regions of the brain process information related to one's own face versus the faces of others.

In one of the earliest studies on the physiology of self-facial recognition, Preilowski (1979) evaluated differences in galvanic skin response (GSR) in "split-brain" patients when they were shown pictures of their own face and those of familiar others. "Split-brain" patients are individuals who have had their corpus callosum, the thick nerve bundle that connects the left and right hemispheres of the brain, severed in order to reduce interhemispheric seizure activity that can occur in serious cases of epilepsy. This radical surgical procedure is rarely performed, and it is used only in the most extreme cases of epileptic seizure that could not otherwise be controlled and were highly disruptive to the individual affected. After surgery, these split-brain patients were given tests to see how well they could identify themselves and others in photographs. Each participant was shown pictures of their own face or the faces of others. Photographs were presented either in the far left field of vision (which is processed in the right hemisphere) or in the far right field of vision (which is processed in the left hemisphere). In split-brain patients, the interconnection between the two hemispheres is severed, and in a sense, each hemisphere (i.e., right and left) now operates independently of one another.

After the surgery, the GSR, specifically the electric charge of each hand, was evaluated as a measure of physiological arousal working under the assumption that increased cerebral activity in a particular hemisphere would be associated with an increased GSR reading in the contralateral hand. Interestingly, it was found that there were no differences between cerebral hemispheres in the GSR measurements when the participants were shown the faces of others. However, when participants were shown pictures of their own face, their left hand (i.e., right hemisphere reaction) had twice the GSR reaction than was measured in their right hand (i.e., left hemisphere reaction). Furthermore, individual's GSR response to seeing their own (self) face was four times greater in their left hand (i.e., right hemisphere reaction) than was their GSR response to seeing the faces of others when measuring the right hand (i.e., left hemisphere reaction) response. Overall, it was concluded that there were significant increases in the GSR when the right hemisphere was shown pictures of one self. Importantly, this was one of the first empirical studies to suggest that the right hemisphere is specific for recognizing one's own face.

A similar study of self-facial recognition was conducted with individuals suffering from intractable (i.e., uncontrollable) epileptic seizures. Individuals with this condition often have to consider having elective neurosurgery to remove the brain tissue where their seizures are originating, with the goal of eliminating their seizures. To minimize the potential neuropsychological complications of such surgeries, patients undergo Wada testing (named after the man who developed the procedure) prior to surgery. During this procedure, individuals have each of their cerebral hemispheres put to sleep, one at a time, to determine which hemisphere primarily controls their language skills. Typically, language capability is located predominately in the left hemisphere for about 90% of individuals. During Wada testing, if an individual is unable to name the images presented in pictures that they are shown, it is concluded that the hemisphere anesthetized at that time is associated with expressive language. This information is crucial as neurosurgeons are hesitant to perform surgeries near the language centers of the brain. Similar to individuals with "split brains," individuals who have Wada procedures offer unique opportunities to determine the specific neuropsychological abilities that are related to each of the cerebral hemispheres.

Keenan, Nelson, O'Connor, and Pascuale-Leone (2001) studied five individuals undergoing Wada testing to determine if the cerebral hemispheres process the faces of the self and of others differently. The authors used a morphing procedure to create pictures that were a combination of the participant's own faces and that of a famous person (with 50% self and 50% famous person). While a specific hemisphere was anesthetized, the participants were then shown the morphed images and asked to remember them. After recovering from the anesthesia of a particular hemisphere, the participants were then asked to choose the pictures they had seen while anesthetized. The results indicated that following anesthesia to the left hemisphere (when the left hemisphere was asleep but the right hemisphere was awake), all five participants were able to consistently remember having seen their own face, but they were unable to recall having seen any of the famous faces. In contrast, following anesthesia to the right hemisphere (when the right hemisphere was asleep and the left hemisphere was awake), four of the five participants were able to consistently recall that they had seen the face of the famous person but they were unable to recognize their own face. These results clearly suggested that the right hemisphere is essential for recognizing one's own face.

There have been numerous other neuroimaging studies of self-facial recognition, and even though methods varied to a minimal extent, they generally have demonstrated similar results. Devue and Bredart (2011) published a review article of these studies and summarized the cerebral differences in neurophysiological activity found when individuals processed images of their *own* face (i.e., self) versus the faces of *familiar others* (i.e., best friends) and *unfamiliar others* (i.e., strangers). They compared data from 41 studies and concluded that although both hemispheres are involved in self-referential processing, the right hemisphere is predominantly associated with self-facial recognition (see Fig. 4). In fact, 33 of the

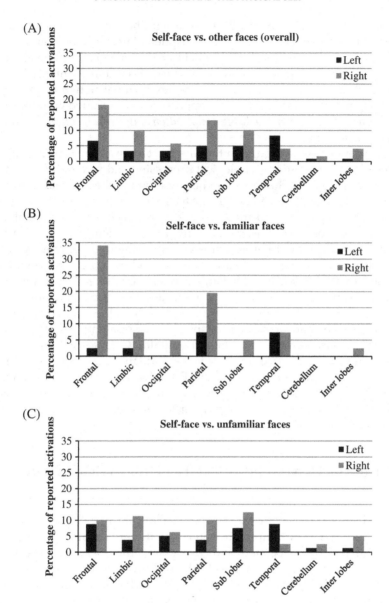

FIG. 4   Percentage of activation reported in cerebral areas in each hemisphere on all activations reported in previous neuroimaging studies: (A) all studies comparing self-face processing with another face processing collapsed, (B) in studies comparing self-face processing with familiar face processing, and (C) in studies comparing self-face processing with unfamiliar face processing (Devue & Bredart, 2011).

41 studies they looked at implicated the right hemisphere, whereas only eight primarily implicated the left hemisphere. Furthermore, their review of the data indicated that the frontal (14 of 15 studies) and parietal lobes (8 of 11 studies) were particularly related to self-facial recognition. Devue and Bredart (2011) concluded: "A complex bilateral network involving frontal, parietal, and occipital areas appears to be associated with self-face recognition, *with a particularly high implication of the right hemisphere*" (p. 40, emphasis added).

The previous studies suggest that the right hemisphere is primarily responsible for the neuropsychological processing and recognition of one's own face. However, this begs the question as to whether the brain has neuropsychological processes specific for processing only one's face or all of one's body parts. To address this question, Meador, Loring, Feinberg, Lee, and Nichols (2000) asked 32 individuals to identify their own left hand, while their right hemispheres were anesthetized. Only eight were able to correctly recognize their own hand, and the other 24 individuals claimed that their left hand belonged to someone else. This study suggested that the right hemisphere is associated with the processing of not only the face but also other physical parts of the self. Other studies have also suggested that the right hemisphere is associated with recognizing one's body parts (Frassinetti, Maini, Romualdi, Galante, & Avanzi, 2008) and also one's own voice (i.e., self-voice; Kaplan, Aziz-Zadeh, Uddin, & Iacoboni, 2008; Nakamura et al., 2001; Rosa, Lassonde, Pinard, Keenan, & Belin, 2008). Increasingly, it appears that the right hemisphere processes self-specific information for many characteristics of the physical body, and not just for the face.

9 RIGHT HEMISPHERE AND THE PSYCHOLOGICAL SELF

In addition to studying the neurological origins of the "physical self," some studies have attempted to determine if there are also specific neurological regions associated with processing the "psychological self." Whereas the physical self relates to the perceived "ownership" of one's unique physical traits, the psychological self has been conceptualized as those personality characteristics and behavioral traits that are unique to each individual (i.e., intelligence, social skills, thoughts, emotions, sense of humor, and coping style). Just as we all have our own physical body, we also have our own unique set of psychological attributes.

To determine the neurological basis of a psychological self, Lou et al. (2004) completed a study to see if the right hemisphere processes information differentially when individuals consider their own personality traits versus those of other individuals. In this study, 13 Danish individuals had PET neuroimaging of their brains taken when resting to determine their baseline levels of neurophysiological functioning. They were then presented with 75 different adjectives commonly used to describe different personality traits and asked to rate how well the adjectives presented described themselves, how well they described their best friend, and how well they described the Danish queen.

Participants were then scanned again, approximately five minutes later, at which time the 75 adjectives were shown, one at a time. The researchers measured the differences between the baseline scans and the intervention PET scans for each word, examining the regional cerebral blood flow for each of the three conditions (i.e., self, best friend, and Danish Queen). The results of the study indicated that "activation increased in the left lateral temporal cortex and *decreased in the right inferior parietal region* with decreasing self-reference" (Lou et al., 2004, p. 6827, emphasis added). In simple terms, the right inferior parietal lobe (RIPL) was most active when describing the personality characteristics of themselves, less active when describing the personality characteristics of close others (i.e., one's best friend), and the least active when describing the personality characteristics of general others with which they had no direct relationship (e.g., the Danish queen). The authors concluded that the RIPL of the brain is related to processing information that is self-referential. In other words, they found that their participant's psychological sense of their own personality attributes (i.e., the psychological self) to be significantly related to heightened activity in the RIPL of the brain, thereby providing another piece of evidence of the importance of the right hemisphere, and especially right parietal lobe, in processing the sense of a psychological self.

10 NEUROTECHNOLOGY AND THE SELF

Most studies conducted on the neurology of the self have been fundamentally limited by the fact that they, like most neuroimaging studies, have been correlational in nature. This means all that can be concluded from these functional neuroimaging studies is that different levels of physiological activity in certain areas of the brain are associated with certain neuropsychological tasks (e.g., medial temporal lobes are related to memory, as per Patient HM; Broca's area in the left frontal lobe is related to expressive language, as per Tan Tan; the RIPLs are related to the psychological sense of self). Most studies to date have not shown the causal relationships that exist between specific cerebral regions and precise self-processing skills. However, two studies have suggested a causal link between the right hemisphere in general and RIPL specifically, in terms of the ability to process one's own face.

Keenan et al. (2001) applied TMS to 10 healthy individuals to determine if increased cortical activity in the right hemisphere was associated with an increased ability to recognize one's own face versus the faces of famous individuals. TMS uses magnetic pulses to either excite or inhibit neurophysiological activity of the specific brain regions that are targeted, which allows for causal relationships to be inferred. In contrast to their previous Wada study in which anesthesia was used to *decrease* right hemisphere activity, in this follow-up study, Keenan and colleagues *increased* the functioning of each hemisphere by applying magnetic stimulation to each hemisphere separately. As they hypothesized, they found that increasing the neurophysiological functioning of the right hemisphere was associated with increased ability to identify one's own face, compared with the faces of famous individuals. In contrast, they also found that increasing the excitability of the left hemisphere did not lead to any differences in recognizing either one's own face or famous faces.

In a similar study, Uddin, Molnar-Szakacs, Zaidel, and Iacoboni (2006) used repetitive transcranial magnetic stimulation (rTMS) to inhibit the functioning of the RIPL in eight healthy individuals to determine the causative relationship between RIPL functioning and the ability to recognize one's own face. In this study, the researchers presented images of the participant's face that were morphed with images of other's faces. Each photograph presented combinations of one's own (self) face and other's faces that were morphed in 20% increments (i.e., 0%, 20%, 40%, 60%, 80%, and 100%), with 100% showing the entire "self" face and 0% showing the entire "other" face. The morphed faces were classified as primarily "self-based" if they contained more than 50% of an individual's own facial features, and as "other-based" if they contained more than 50% of another's facial features. Participants had twenty minutes of rTMS applied to their RIPL to inhibit neuropsychological processing in this area, and then, they were subsequently reassessed to see how accurately they could recall the "self" or "other" facial images. The results indicated that when a person's RIPL functioning was inhibited, they were less able to recall the pictures that were primarily composed of their own facial features. The researchers also applied rTMS to disrupt the functioning of the left inferior parietal lobe as a control. Here, they found that there was no significant effect on a person's ability to discriminate between their own face and another's face. Taken together, these studies suggest that there is a causal relationship between RIPL functioning and one's self-facial recognition, again suggesting the predominance of the right parietal lobe in processing a sense of the physical self.

However, TMS has also been used to show a causative relationship between the right hemisphere and the processing of the psychological self. In extending their PET study of Danish individuals, Lou et al. (2004) examined how well 25 American research participants described themselves versus their best friends, when TMS was applied over the parietal cortex in order to selectively disturb neuronal circuitry in this region of the brain. With TMS applied, they found a decrease in the ability to retrieve information about their own mental self compared with that of others (i.e., best friend). Overall, they found that "[t]he right inferior parietal lobe (RIPL) was particularly active during retrieval of self-referential information" (p. 6831). Other studies on physical first-person perspectives including position in space, imagining agency, and body representation have also shown significant activation in the RIPL (see Lou et al., 2004 for additional references). As Lou and associates point out, collectively, all these studies "point to a role for the right lateral parietal region in representation of the physical self. With our present results showing that this is also the case for the mental self, we conclude that the right inferior parietal cortex is selectively activated in self-representation in general" (p. 6831).

11 CONCLUSION: SELF AS PROCESS

> To put it another way, given that the self is a process and not a substantial thing, it may be possible to shut down this process under certain conditions and then start it up again. *(Thompson, 2015, p. 362)*

Neurological research suggests that the right hemisphere and RIPL are primarily responsible for integrating sensory and mental experiences to create a sense of the self, including the physical self (e.g., the face, hand, body, and voice) and the psychological self (i.e., mental characteristics). Case studies of individuals without the ability to encode memories (e.g., patient HM) are examples of disorders of the autobiographical self (although we conceptualize this as primarily a disorder of memory). All of these studies suggest that the self is not best conceptualized as a permanent entity, but as the experience of complex integrated neuropsychological processes that create the sense of self.

A review article by Decety and Sommerville (2003) concisely summarizes the nature of this research by stating, "the right inferior parietal cortex in conjunction with prefrontal cortex may be critical in distinguishing the self from the other" (p. 532). Rather than considering these brain areas as individual modules, they view them as a collection of interconnected regions that are essential for the subjective experience of a "self." However, they offered the caveat that the role of each specific region in self-facial processing remains highly hypothetical.

Research in developmental psychology has also indicated that the brain develops neuropsychological processes over time that are specific to creating and recognizing a "self." For example, around the same time that William James proposed that the self was nothing more than passing states of consciousness, Preyer (1889) recognized that children develop a sense of self over time (that the ability to create a sense of self is a neuro-developmental process), stating the following:

> The behavior of the child toward his image in the glass shows unmistakably the gradual growth of consciousness of self out of a condition in which objective and sub-jective changes are not yet distinguished from each other. *(quoted in Molnar-Szakacs & Uddin, 2013, p. 173)*

One hundred years later, neurodevelopmental research confirmed that infants start to recognize themselves in the mirror between the ages of 18 and 24 months (Lewis, Sullivan, Stanger, & Weiss, 1989). These observations suggested that a "self" does not necessarily exist at birth but rather that it is a neuropsychological process that develops as the brain matures.

These studies suggest that, rather than conceptualizing the self as a permanent entity, it may be best conceptualized as a series of integrated neuropsychological processes that change continually depending on different causes and conditions. Evan Thompson, a Buddhist philosopher who has written about the nature of what Buddhism would regard as the dependent self, understands the self not as an entity but rather as a process. Discussing the insights of Nagarjuna, one of the most renowned philosophers of Buddhism, Thompson says:

> Our ordinary or everyday concept of the self is the concept of a subject of experience and an agent of action, not of an inner and substantial essence of the person. Furthermore, when we look carefully at what we apply our ordinary concept of self to in the world of our individual and collective experience, we don't find any inherently existent thing or independent entity; what we find is a collection of interrelated processes, some bodily or physical, some mental or psychological ... More simply stated, *the self isn't a thing or an entity; it's a process. (Thompson, 2015, p. 323, emphasis added)*

As the current and previous chapters illustrate, both case studies and empirical research suggest that the "sense of self" can be altered through injury or disease (e.g., TBI, tumor, and stroke) or scientific interventions (e.g., anesthesia and TMS/rTMS). When the self is conceptualized as the experience of integrated sensory experiences, this suggests that the process of creating a sense of self may also be enhanced or inhibited through behavioral practices, much the same way that neural networks associated with musical skills can be influenced by practice and repetition. This raises the question as to whether or not religious practices such as prayer or meditation may influence the experience of the self (or, conceptualized differently, the experience of selflessness). Interestingly, as the next chapter will demonstrate, empirical research is beginning to show relationships among the right hemisphere, selflessness, and spiritual transcendence.

5

The Neuropsychology of Spiritual Transcendence

> *To lose yourself, as if you no longer existed, to cease completely to experience yourself, to reduce yourself to nothing is not a human sentiment but a divine experience.* **Saint Bernard of Clairvaux (1090–1153 CE) (from De diligendo Deo x.27, On loving God, 29; Opera omnia, i. 990; quoted in** *Kilcrease, 2014, p. 270)*

The previous chapters suggest that there is no absolute, unchanging "self," but rather, a "sense of self" is formed through multiple neuropsychological processes that create physical, psychological, and autobiographical selves that change over time and circumstance. This contention is supported in part by research that indicates that injuries to those parts of the brain associated with self-orientation (e.g., right hemisphere/parietal lobe) result in "disorders of the self" (e.g., left-sided neglect, anosognosia, and asomatognosia). However, in addition to conceptualizing a lessened

© 2019 Elsevier Inc. All rights reserved.

sense of self as a "disorder," it may also be important to determine if there are positive behavioral attributes associated with a decreased sense of the self. Or stated in another way, it may be important to determine the characteristics of individuals who have increased experiences of selflessness. In fact, recent neurological research of spiritual experiences suggests that a reduced focus on the self, generated through either ritualistic practices or certain brain disorders associated primarily with the right hemisphere/parietal lobe, can lead to the transcendent spiritual experiences similar to those reported by mystics throughout history.

1 SELFLESSNESS AND TRANSCENDENCE: AN EXPLORATION

Research over the past several decades has, in fact, suggested that individuals can reduce focus on the self, primarily through religious rituals and practices (i.e., turn "off" switch to the self, as per Jonathan Haidt, 2012), which leads to spiritually transcendent states. With the advent of neuroimaging techniques and particularly over the past 25 years, the neurological study of spiritual and religious experiences has expanded significantly and facilitated a better understanding of the neurological nature of transcendence. As expected, the initial studies in this area were simplistic in terms of neurological methodologies and definitions and measures of spiritual constructs. Although these types of studies have significantly improved over time, many of these weaknesses continue to exist. As such, in this chapter, we focus on creating an understanding of spiritual experiences that are specifically related to the neuropsychological process of selflessness. In general, we define these selfless experiences as forms of "spiritual transcendence" that involve the sensation of a greatly decreased sense of self and an associated intense emotional connection to a higher power, the cosmos, nature, or however this connection may be understood (for more on this, see Chapter 6). We find that it may be possible to explore such important human experiences through both subjective and empirical means. With this noted, we acknowledge the inherent difficulties and contradictions associated with attempting to describe spiritual experiences that are ineffable and therefore in many ways remain indescribable.

2 WHAT THE HECK IS GOING ON?

To start off, we describe the circuitous route we took in our decade long investigation of the neuropsychological foundations of spirituality. Our path illustrates the necessity of having diverse disciplines from the sciences and humanities challenging each other to better understand the nature of spiritual experiences. The sciences provide the objective standards and methods necessary to validate empirical hypotheses, while the humanities provide the collective wisdoms across history to improve our understanding of the nature of humanity and our relationship to the divine.

In general, scientists conduct methodologically rigorous studies to prove or disprove their hypotheses and theories of interest, whatever they may be. When lucky, the data support the models, and the researchers look quite intelligent. At other times, researchers stumble upon data that do not make much sense, and only after conferring with individuals with different perspectives and careful exploration of possible causes do the answers become clear. Following is an example of how a collective group of researchers at the University of Missouri (MU) came to explore the existence of an association between right parietal lobe (RPL)-related selflessness and spiritual transcendence in groups of individuals with traumatic brain injury (TBI). In all honesty, it was more of the stumble-upon-some-interesting-data situation than confirmation of a well-reasoned, a priori hypothesis.

MU was awarded a grant from the Pew Charitable Trusts to start a *Center for the Study of Religions and the Professions*. The goal of the center was to pull together a team of scholars from multiple academic disciplines to discuss the many ways in which religion was relevant to the professions and particularly as related to the public. Attendees came from medicine, philosophy, law, social work, journalism, psychology, business, and religious studies. It was a wonderful opportunity to work with scholars from outside one's area of expertise to learn of the many ways that religions influence what we do in our work and daily lives.

As part of our responsibilities to the grant, we were required to conduct collaborative research related to religion and our individual disciplines. Using experience in clinical neuropsychology as a springboard, a team of researchers was engaged to conduct a study to determine if specific cognitive abilities (e.g., memory, language, spatial skills, attention, and abstract reasoning) are associated with "spiritual experience" (at the time we did not provide any more specific definition). If we had cultivated any hypothesis at the time, it was that an increased ability to engage in abstract reasoning or increased ability to focus attention would be related to increased spirituality. We came to propose that in order to have complex and abstract neuropsychological experiences associated with a sense of unity with higher powers or the divine, one must have relatively strong abilities to focus attention and to reason abstractly. Confining ourselves for the most part to the discipline of neuropsychology and the tools it uses (i.e., tests of memory, attention, and language), we narrowly conceptualized the awe- and wonder-filled mystical experience of relating to the divine as a combination of relatively simple cognitive processes.

II. THE "SELF" AND SELFLESSNESS

Consistent with the lesion analytic studies previously described in Chapter 3, we studied 26 individuals who were referred for neuropsychological evaluation of TBI (Johnstone & Glass, 2008). We worked under the assumption that if spiritual experiences are related to different cerebral regions and different neuropsychological processes, then it would be easier to identify such brain-spirituality relationships in persons with brain dysfunction in which there is greater heterogeneity of cognitive abilities. We measured spirituality with the Index of Core Spiritual Experience (Inspirit) scale (Kass, Friedman, Lesserman, Zutterman, & Benson, 1991), which assesses general spiritual experiences and one's sense of transcendence with questions such as the following:

- How often have you felt as though you were very close to a powerful spiritual force that seemed to lift you outside yourself?
- How close do you feel to God?
- Have you ever experienced God's energy or presence?
- Have you ever experienced a feeling of unity with the earth and all living beings?

The participants also completed standard neuropsychological evaluations that assess cognitive abilities (i.e., intelligence, memory, attention, language, visual-spatial skills, and executive functions) that are associated with different regions of the brain. The following cognitive abilities were measured with the following standardized neuropsychological tests:

- Verbal memory (Wechsler Memory Scale-III (WMS-III) Logical Memory (LM) I subtest; Wechsler, 1997)
- Visual memory (WMS-III Visual Reproduction (VR) I subtest; Wechsler, 1997)
- Divided attention (Trail Making Test Part B; Reitan, 1992)
- Expressive language (Controlled Oral Word Association Test; Benton & Hamsher, 1989)
- Spatial perception (Judgment of Line Orientation (JOLO) Test (Benton, Hamsher, Varney, & Spreen, 1983)

We conducted correlation analyses between the Inspirit and the different neuropsychological tests (see Table 1). In Table 1 (and the following tables in this chapter), correlations show how performance on a particular neuropsychological test is associated with spiritual transcendence measured by the Inspirit and other measures of spirituality. To ease interpretation of the data, the correlations are presented so that (a) positive correlations indicate that higher levels of spirituality are associated with better neuropsychological performance and (b) negative correlations indicate that higher levels of spirituality are associated with worse neuropsychological performance.

Much to our surprise, in this first study, we found that the only neuropsychological measure that significantly correlated with spiritual

TABLE 1   Correlations Between Spirituality and Neuropsychological Indexes**

Neuropsychological indexes	Inspirit
RIGHT HEMISPHERE	
Right parietal lobe	
JOLO	−0.56**
Right temporal lobe	
WMS-III VR I	−0.27
LEFT HEMISPHERE	
Left temporal lobe	
WMS-III Logical Memory I	0.13
Left frontal lobe	
COWAT Total	0.00
General frontal lobe	
Trail Making Test Part B	−0.09

** $p < 0.01$; Inspirit, Index of Core Spiritual Experiences; COWAT, Controlled Oral Word Association Test; JOLO, Judgment of Line Orientation; WMS-III, Wechsler Memory Scale-III; LM, Logical Memory subtest; VR, Visual Reproduction subtest.

transcendence was the JOLO test, which measures a person's ability to estimate spatial relationships. However, confounding these results even further was that the data analysis indicated that the JOLO was *negatively* related to spirituality measured by the Inspirit test and to a fairly large degree. This result was unexpected, and our first response was "what the heck is going on?"

In order to better appreciate these findings, it is helpful to understand more about the JOLO test. Fig. 1 shows four cards with the stimuli used in the JOLO test. In each example, the lowermost portion of each card shows the spokes of half a wheel that are numbered from 1 to 11. Each card has two spokes (or lines) in its upper portion, with each of these lines corresponding to the spatial position of one of the 11 spokes shown on the bottom portion. In the test, individuals are to determine which of the 11 lines of the lower portion corresponds spatially to the two lines presented in the upper portion. The upper left example in Fig. 1 has two lines on the upper portion of the card that correspond to the lines numbered 5 and 10 on the lower portion with the 11 spokes. In the JOLO test, a total of 30 different stimulus cards are presented with a variety of different patterns of two lines, and a total score is obtained by counting up the total number of cards that a person has answered correctly. The higher the score one receives, the better their ability to perceive spatial relationships among objects.

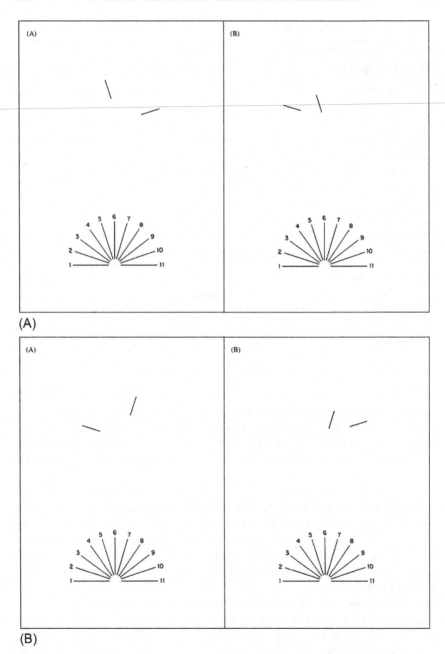

FIG. 1 The JOLO test.

When reviewing our results, we first tried to understand why *decreased* spatial perception (i.e., lower scores on the JOLO test) would be related to *increased* spiritual transcendence (i.e., receiving higher scores on the Inspirit scale). How could a reduced ability to estimate spatial relationships be related to more self-reported closeness to the divine? At first, we reasoned that this finding could be associated with the altered sense of space reported by people who have spiritual experiences (as has been concluded by other researchers, e.g., Newberg, Alavi, Baime, Mozley, & d'Aquili, 1997; Newberg, Pourdehnad, Alavi, & d'Aquili, 2003), but we came to believe that this was too simplistic. After much discussion, we could not come up with a plausible explanation.

In retrospect, we made an error that is common to most disciplines, namely, that we study concepts in terms and with tools that are specific to our own disciplines (in other words, we find what we seek, and we seek what we know). As neuropsychologists, we investigated spirituality as a primary *cognitive* construct. That is, we conceived of spiritual experiences as being primarily associated with thinking skills. But descriptions of the experiences of mystics throughout history clearly indicate that spiritual transcendence is primarily related to an emotional connection with the divine and not so much how one cognitively understands or intellectually relates to an abstract construct such as "God" or the divine.

We questioned whether the significant relationship between the Inspirit and the JOLO results might be reflective of the fact that individuals with lower IQ scores (and cognitive abilities) have been shown to report higher levels of spirituality (Lewis, Ritchie, & Bates, 2011; Lynn, Harvey, & Nyborg, 2009). Our results did indicate that lower intelligence was associated with higher levels of spirituality, but this correlation was not statistically significant. We also questioned if increasing severity of TBI could account for the increased spirituality. In other words, did greater neuropsychological impairment and overall lower neuropsychological abilities correlate with higher Inspirit scores? Those data did not support this as we noted that the neuropsychological measures generally associated with right-hemisphere abilities (i.e., visual memory and spatial perception) were negatively associated with increased spirituality, while measures generally associated with left-hemisphere abilities (i.e., verbal memory and expressive language) were positively associated with increased spirituality. These results suggested that increased spirituality was related to *decreased* functioning in the right hemisphere but *increased* functioning in the left hemisphere (a pattern to remember when reviewing our other studies that are described below). Overall, the results indicated that increased connection with a higher power was not related to globally impaired cognition, but rather that it was related to a combination of increased and decreased neuropsychological abilities across different regions of the brain.

We eventually determined that we needed to look at the data from a different perspective. After initially conceptualizing the JOLO test as a measure of spatial perception, we decided to conceptualize it as an index of the functional integrity of the RPL. This seemed plausible as the JOLO (and other tests of spatial perception) have consistently been shown to be related to functioning of the RPL (Benton et al., 1983). A study of individuals from a registry of individuals with specifically located (i.e., focal) neurological lesions confirmed that lower performance on the JOLO is primarily associated with lesions in the posterior RPL (Tranel, Vianna, Manzel, Damasio, & Grabowski, 2009). This study assisted in putting "new empirical teeth in the localizing value" of the JOLO for RPL functioning (Tranel et al., 2009, p. 219). In essence, these studies suggested that the JOLO test could be used as a measure to determine the functional integrity of the RPL (i.e., how well the RPL was functioning).

Considering the data in this manner, we inferred that if the RPL was shown to have reduced functioning (as per the JOLO test), it was likely that other neuropsychological abilities associated with the RPL may also have reduced functioning. In general, the RPL has been consistently shown to be associated with spatial perception (e.g., JOLO), left-sided tactile sensitivity (i.e., the ability to feel things on the left side of the body), and self-orientation (see Chapters 3 and 4). When conceptualized together, these related outcomes could be interpreted to suggest that reduced performance on the JOLO test was reflective both of weak RPL functioning and weaknesses in self-orientation. Putting all of this information together, we came to the conclusion that a reduced ability to focus on the self could lead to increased connection to things beyond the self, facilitating experiences of transcendence. Although the inference was simple, when we conceptualized the data in this manner, the results we had obtained started to make a lot more sense.

3 NEUROPHYSIOLOGICAL STUDIES OF TRANSCENDENT EXPERIENCES

After realizing that we really did not know much about the neural foundations of spiritual experiences, and particularly how reduced functioning of the RPL may be related to increased spirituality, we conducted literature reviews to see if any other studies suggested a relationship between reduced RPL activity and increased spirituality. Much to our surprise, we found several functional neuroimaging studies that seemed to support our findings. The studies focused on blood flow through the brain (i.e., cerebral perfusion) during spiritual/religious activities (e.g., during meditation or prayer), and they generally showed that the parietal lobes and particularly the RPL had decreased activity (based on reduced blood flow) during these practices. We naively expected these types of religious ritual practices to be associated with increased cerebral activity and assumed that transcendent experiences (commonly described in terms of awe, wonder, bliss, selflessness, and unity) would be associated with increased neural activity throughout the entire brain. We did not expect that these exceptional experiences would actually involve inhibition of certain neural networks. Why would particular regions of the brain be turned off during spiritual experiences? Initially, it just did not make sense ... unless it was the focus on the self that was being diminished.

Two of the first neuroimaging studies we reviewed investigated the cerebral activity of experienced spiritual practitioners during different religious rituals. Although the studies are limited methodologically, they provided important information for our new hypotheses. Specifically, Andrew Newberg and his colleagues at the University of Pennsylvania conducted neuroimaging studies of the regional cerebral blood flow (rCBF) of individuals engaged in religious practices. These studies attempted to determine if different regions of the brain show altered levels of physiological activity when individuals engaged in spiritual practices that form part of their religious traditions. Single-photon emission computed tomography (SPECT) imaging was used to identify areas of increased and decreased blood flow in the brain during concentration meditation for long-term Buddhist practitioners and later for centering prayer performed by Franciscan nuns who had decades of experience with this type of prayer ritual. Although there are methodological limitations to these studies (e.g., sample sizes of eight and three, respectively, multiple statistical analyses, numerous cerebral regions of interest), they nonetheless provided interesting insights into the nature of spiritual transcendence and also showed reduced physiological activity of the RPL.

In the first study, Newberg and his colleagues (1997, 2001) evaluated regional blood flow before and after the meditative practice of eight proficient Buddhist meditators. The participants engaged in meditation during which they visualized an object as part of their spiritual practice. Analysis indicated that following their meditation practice, the Buddhist meditators had increased blood flow primarily in the frontal cortex (including the cingulate gyrus, inferior and orbitofrontal cortex, and dorsolateral prefrontal cortex (DLPFC)) and in the thalamus located at the top of the brain stem. Additional analyses indicated that the greatest increase in blood flow following meditation occurred in the orbitofrontal region (a 25.8% increase) but that at the same time, there was an associated *decrease* in blood flow (although it was not highly significant) in only one area of the brain, namely, the superior (i.e., upper portion of the) parietal lobes (a 5.0% decrease). Similarly, other analyses indicated increased blood flow in the right side of the thalamus but *decreased* blood flow in the superior parietal, right lateral temporal, and left inferior (i.e., lower portion of) temporal lobes. In general, the results indicated that Buddhist meditation was associated with increased physiological activity primarily in the frontal lobes and with an associated *decrease* in activity in the superior parietal lobes. Although hemispheric differences were not found, the results nonetheless indicated that physiological activity in the parietal lobes *decreased* during meditation. The authors concluded that the increase in frontal lobe activity may be related to the intense attention that is necessary for meditative practices and that the decrease in parietal lobe activity might be related to a diminished sense of spatial awareness often reported by meditators—although we suggested this may be more appropriately interpreted as a reduction in self-processing, rather than a reduction in the sense of space. We note that the "self" is often defined in terms of space, time, and relationship (i.e., I am this body in this space, at this time, in relationship with these objects), so a reduction in one's sense of space could also be related to a reduction in the overall sense of self.

Newberg et al. (2003) conducted a similar SPECT study of Franciscan nuns engaged in centering prayer and again saw both increased frontal activity and reduced parietal lobe activity during the nun's prayer rituals. In general, their research addressed whether the regional blood flow patterns observed during the "spiritual" practices of Buddhist meditators engaged in a visually based meditation were the same as those experienced during the "religious" practices of Christian nuns engaged in a verbally based prayer. It was hypothesized that if the blood flow patterns are the same, this would suggest that the spiritual experiences of individuals from quite differently oriented faith traditions (nontheistic vs. theistic) have common neuropsychological foundations (e.g., decreased self-orientation), which are interpreted differently based on the cultural and religious orientations and outlooks of the ritual practitioners.

Overall, the results of this second study with Franciscan nuns did show results that were generally similar to those of the Buddhist meditators in terms of reduced parietal blood flow. Following approximately 50 minutes of centering prayer, blood flow increased 7.1% in the prefrontal cortex, 9.0% in the inferior frontal lobes, and 6.8% in the inferior parietal lobes (IPLs). Most importantly, the researchers found a "strong inverse correlation" between the blood flow in the prefrontal cortex and the (ipsilateral superior) parietal lobe (in research notation, $r = -0.75; p < 0.01$). While it is difficult to determine how such a statistically significant result could be obtained in a sample of only three individuals, nevertheless these studies did suggest that increased neural activity of the frontal lobes and *decreased* activity of the parietal lobes were associated with both Buddhist meditation and Christian centering prayer. Of note, the authors concluded on the basis of descriptions given by the Buddhist meditators and the Franciscan nuns that the spiritual experiences each achieved as a result of their respective practices involved an increasing sense of a universal connectedness, a greater sense of unity over diversity, and *a decreasing sense of the self.*

We were intrigued to learn that the participants in both of these studies showed decreased parietal lobe functioning during their respective spiritual practices. For the Buddhists using meditation, they said they attained a momentary glimpse of true reality, while the nuns using centering prayer reported that they achieved communion with God. Whether Buddhist or Catholic, the participants had reduced parietal lobe functioning and reported increased feelings of connection with the cosmos by losing their false sense of self, and either glimpsing true reality (Buddhists), or as a deep experience of union and direct connection with God (Christians). Although the participants had achieved certain similar patterns of CBF, they conceptualized their experiences in different ways based upon their particular systems of belief. We believed that our initial study with persons with TBI, when considered together with Newberg's SPECT studies, suggested that there may be a universal neuropsychological process (i.e., reduced RPL activity and related selflessness) that is regularly associated with experiences of spiritual transcendence (e.g., connecting with a higher power beyond the self or momentarily glimpsing true reality) for all individuals regardless of religious belief.

Looking further into the literature, we found several other neuroimaging and meditation studies that tended to indicate that the functioning of the parietal lobes is inhibited during meditation. For example, one of the earliest neurophysiological studies of spiritual practices evaluated differences in regional glucose metabolism (anterior vs. posterior brain regions; right vs. left cerebral hemispheres) in the brains of eight experienced individuals engaging in a yoga relaxation meditation (Herzog et al., 1990). The study reported generally nonsignificant results due to intersubject variability, but the results provided interesting suggestions for different patterns of regional brain activation and inhibition during these practices. Specifically, the authors reported remarkable differences in specific regions of the brain, with all participants showing increased anterior activity (i.e., in the front part of the brain) but *reduced* posterior (i.e., in the rear part of the brain) CBF in the occipital and parietal lobes during the meditation. Although no conclusions were made regarding a relationship between diminished posterior brain functioning and a decreased sense of the self, it was noted that the participants generally reported the "feeling of relaxation, peace, and detachment between body and soul" (p. 184). These studies were suggesting increasingly that our focus on reduced RPL functioning and increased spirituality might be an accurate appraisal of what was occurring.

A more recent literature review on the neurophysiology of spiritual and meditative practices generally supports the results of these early neuroimaging studies. Researchers in Australia conducted a review of 96 neuroimaging studies to determine if differences exist in the balance of cerebral activity between parietal and prefrontal regions between secular mindfulness meditation and religious/spiritual practices (Barnby, Bailey, Chambers, & Fitzgerald, 2015). The researchers investigated changes in cerebral activation and inhibition for three groups engaged in distinct but related practices (i.e., mindfulness meditation, spiritual practice, or a combination of both). Mindfulness was defined as involving the expansion of attention in a nonjudgmental and nonreactive manner in order to become more aware of one's current sensory, mental, and emotional experiences. Spiritual practices were defined as involving a focus on a sense of connection to a higher entity. Their review concluded that the "evidence seems to support that the balance of *inhibited* parietal activity and amplified prefrontal activity may lead to the strengthening of the religious/spiritual experience" (p. 222, emphasis added). The reviewers suggested that spiritual practices may influence the activity or inactivity of specific regions of the brain, for example, inactivity in parietal cortex, which is associated with distinguishing between the "self" and "others." Summarizing their research on the relationship between spiritual belief and the parietal cortex, they indicate the following:

> A relative decrease in activation of the parietal cortex, specifically the inferior parietal cortex, appears to be reflective of spiritual belief, whether in the context of meditation or not. *(Barnby et al., 2015, p. 219)*

4 "ROUND TWO"

Encouraged by what we learned from the literature and our earlier research results, we attempted to replicate the findings of our first study in a subsequent study of 20 individuals with TBI (Johnstone, Bodling, Cohen, Christ, & Wegrzyn, 2012). We wanted to determine if we were on to something interesting and important or if we had only discovered a random association between a measure of spatial perception and spirituality. Once again, we evaluated relationships between the functional integrity of different cerebral regions (using various neuropsychological tests) and self-reported measures of spiritual experiences. In addition to assuming that the JOLO test could measure the functional integrity of the RPL, we used the following measures as indicative of the functional integrity of other general cerebral regions as follows: verbal memory (left temporal lobe), visual memory (right temporal lobe), expressive language (left frontal lobe), divided attention (frontal lobe) (for a review see Lezak, Howieson, & Loring, 2004).

However, to validate our conclusions, we added another measure of spirituality and another measure of RPL functioning. We hypothesized that if decreased RPL functioning was related to one measure of spirituality (i.e., Inspirit), then it should also be significantly related to other measures of spirituality. As a result, we also included the Brief Multidimensional Measure of Religiousness/Spirituality (BMMRS; Fetzer Institute and National Institute on Aging Working Group, 1999), which differentiated between religion (expressed as culturally based behaviors such as prayer, rituals, and service attendance) and spirituality (seen more as an emotionally based connection with the divine, however so conceived). In contrast to the Inspirit that provided one global spirituality score, the BMMRS included measures of multiple dimensions of spirituality including daily spiritual experiences, values/beliefs, meaning, forgiveness, and religious/spiritual coping. Following are several examples of spiritual questions from the BMMRS:

- I feel the presence of a higher power.
- I desire to be closer to or in union with a higher power.
- I feel the love of a higher power for me, directly or through others.
- I am spiritually touched by the beauty of creation.

In order to further validate our findings, we also added another measure of RPL functioning. We hypothesized that if increased spirituality was related to one measure of the functional integrity of the RPL (i.e., the JOLO test), then it should also be related to other measures of RPL functioning. As such, we also administered the left hand finger agnosia test. This test measures the ability of individuals to identify, with their eyes closed, which of the five fingers of their left hand that are touched by an examiner. Being touched on the left side of the body is perceived in the anterior portion of the RPL (Lezak et al., 2004). If this region of the RPL is not functional, if it is injured or inhibited, then individuals will have difficulties feeling when their left fingers are touched as is seen, for example, in individuals with RPL strokes who cannot feel their left arms/hands/fingers. It was hypothesized that if diminished RPL functioning was associated with increased spiritual transcendence, then we would see significant correlations between multiple measures of RPL functioning, for example, between the JOLO and left hand finger agnosia tests and the different measures of spirituality, namely, the Inspirit and the BMMRS.

In this second study, we included additional measures of neuropsychological abilities associated with various brain regions. Rather than focus on these tests in this discussion, we ask (statistically oriented) readers to review Table 2 for an overview that shows the manner by which different cerebral regions are generally associated with spiritual experiences and gives information on the numerical correlations. Overall, consistent with our first study (Johnstone & Glass, 2008), the results showed that right-hemisphere regions appear to be negatively associated with spirituality due to decreased activity, while left-hemisphere abilities appear positively associated with spirituality and are more active.

Our conclusions were also generally supported by the additional measures of RPL functioning that were incorporated. For example,

TABLE 2 Correlations Between Spirituality and Neuropsychological Indexes of Cerebral Functioning

Neuropsychological indexes	DSE	Meaning	Values/ belief	Forgiveness	RS coping	Inspirit
RIGHT HEMISPHERE						
Right parietal lobe						
JOLO	−0.34	−0.02	−0.38	−0.44*	−0.30	−0.60**
Left hand finger agnosia	−0.29	−0.35	−0.32	−0.30	−0.15	−0.34
Right temporal lobe						
BVMTR total	−0.04	0.02	−0.03	−0.13	0.12	−0.23
LEFT HEMISPHERE						
Left parietal lobe						
Right hand finger agnosia	0.13	0.00	0.20	0.27	0.29	0.12
Left temporal lobe						
WMS-III Logical Memory	0.41	0.30	0.23	0.21	0.28	0.21
Left frontal lobe						
DKEFS Letter Fluency	0.17	−0.11	0.17	0.26	0.14	0.13
GENERAL FRONTAL LOBE						
Trail Making Test Part B	0.40	0.20	0.18	0.21	0.29	0.43

*$p < 0.05$; **$p < 0.01$; BMMRS, Brief Multidimensional Measure of Religiousness/Spirituality; DSE, Daily Spiritual Experience; RS, Religious/Spiritual coping, Inspirit, Index of Core Spiritual Experiences; DKEFS, Delis-Kaplan Executive Functioning Scale; JOLO, Judgment of Line Orientation; BVMTR, Benton Visual Memory Test Revised. For more specific information about the neuropsychological tests and their relationships to specific cerebral regions, please see the individual studies (Johnstone & Glass, 2008; Johnstone et al., 2012).

reduced RPL functioning as measured by the left hand finger agnosia test approached statistical significance with the Inspirit measure and all five BMMRS spiritual scales. The findings indicated that the more difficulty individuals had perceiving touch to the left hand, the greater their self-reported spiritual transcendence, meaning that those who had more dysfunctional RPLs exhibited higher levels of self-reported spirituality. The results could be interpreted in several ways. Decreased ability to perceive touch on the left hand could be associated with increased spirituality, although it is difficult to understand directly why this may be the case. However, reduced functioning of the RPL (as inferred from weaker performance on both the JOLO and the left hand finger agnosia test) could suggest decreased functioning of the RPL in general, a lessened ability to focus on the self, and a subsequent increased capacity to connect with things beyond the self. We believed this alternative interpretation made more conceptual and intuitive sense.

Consistent with the first study, no neuropsychological indexes other than the JOLO were significantly associated with measures of spirituality. In addition, also consistent with the first study, several other neuropsychological measures of left-hemisphere abilities were positively associated with spirituality, suggesting that increased spirituality was not reflective of a general decline in neuropsychological abilities common to cases of persons with TBI. Our second study tended to confirm our belief that decreased functioning of the RPL was associated with a decreased sense of self and the subsequent potential for increased experiences of spiritual transcendence. We thought we were on to something but still needed further validation.

5 "ROUND THREE"

Whereas our second study (Johnstone et al., 2012) validated our original findings (Johnstone & Glass, 2008) and extended them by supporting the initial model with additional measures of spirituality (i.e., Inspirit and BMMRS) and RPL functioning (i.e., JOLO and left finger agnosia), our third study attempted to replicate the results by extending them to a population with a different neurological condition. Specifically, we completed a similar study with individuals with seizure disorders rather than TBI (Johnstone, Bayan, et al., 2014). We believed this might be productive, particularly given the purported relationships between epilepsy and increased religiosity (Bear & Fedio, 1979; Greyson, Broshek, Kerr, & Fountain, 2015). Seizures are characterized by abnormal discharges of electric activity within the brain, which can lead to long-term neuropsychological weaknesses with increasing frequency and severity of seizures. We hypothesized that individuals with increased damage to the RPL (which varies depending on the focal site, severity, and duration of the seizures) would experience a decreased sense of self and therefore be more likely to experience increased spiritual transcendence and a greater willingness to forgive. The statistical results we obtained are summarized in Table 3.

Drawing on a sample of 23 individuals with a range of seizure disorders (i.e., generalized, complex and simple partial, and mixed) referred for neuropsychological evaluation, we focused on relationships between measures of RPL functioning (i.e., JOLO and left hand finger agnosia tests) and the BMMRS Forgiveness scale, given that the previous study found that it was the only BMMRS spiritual scale that was significantly correlated with the JOLO. The results (see Table 3) were generally consistent with our second study, in that the BMMRS Forgiveness scale was significantly and negatively

TABLE 3   Correlations Between Spirituality and Neuropsychological Indexes of Cerebral Functioning

Neuropsychological indexes	BMMRS forgiveness
RIGHT HEMISPHERE	
Right parietal lobe	
JOLO	−0.41**
Left hand finger agnosia	0.00
Right temporal lobe	
BVMTR	−0.27
Right frontal lobe	
Left hand finger tapping	−0.24
LEFT HEMISPHERE	
Left parietal lobe	
Right hand finger agnosia	−0.38
Left temporal lobe	
WMS-III Logical Memory I	0.39
Left frontal lobe	
DKEFS Verbal Fluency	−0.25
Right Hand Finger Tapping	−0.37
GENERAL FRONTAL LOBE	
Trail Making Test Part B	−0.49**

*$p < 0.05$; **$p < 0.01$; *BMMRS*, Brief Multidimensional Measure of Religiousness/Spirituality; *DKEFS*, Delis-Kaplan Executive Functioning Scale; *JOLO*, Judgment of Line Orientation; *BVMTR*, Benton Visual Memory Test Revised; *WMS-III*, Wechsler Memory Scale-III. See Johnstone, Bayan, et al. (2014) for specific information about the neuropsychological tests used and their relationships to specific cerebral regions.

correlated with measures of the RPL functionality, meaning that decreased RPL functioning was associated with an increased willingness to forgive. Generally consistent with the previous studies, the Forgiveness scale was also negatively correlated with other measures of right-hemisphere functionality and positively correlated with functional measures of the left temporal lobe. The fact that all measures of frontal lobe functioning were negatively correlated with willingness to forgive was hypothesized to be related to a decreased ability to ruminate about the perceived wrong to the self (i.e., a reduced ability to focus associated with frontal lobe injury).

6 "ROUND FOUR: CROSS CULTURAL/RELIGIOUS EXPLORATIONS"

At this point, our ongoing studies suggested that reduced RPL functioning was associated with increased spiritual experiences. It was still possible that we had only discovered random associations between the JOLO and measures of spirituality, particularly given our small sample sizes. However, we also noted that all three of our studies indicated that the correlations between RPL functioning and spirituality measures were consistently negative and relatively high. We decided to conduct a larger-scale study with more participants, which expanded our investigation to include more diverse populations. Our first studies had been conducted almost exclusively with Caucasian Christians from the United States (as was the case for most neurology of spirituality studies at that time). We thought it would be valuable and most interesting if our next study could validate our neuropsychological model of spiritual transcendence by surveying individuals from different ethnic, cultural, religious, and linguistic backgrounds.

We hypothesized that perhaps self-orientation is a universal neuropsychological process associated with the RPL for all humans, regardless of culture, language, gender, ethnicity, or religious identity. Expanding the hypothesis, we asserted that reduced RPL functioning should be associated with decreased self-orientation (i.e., increased selflessness) for all individuals and that this increased selflessness should be associated with greater spiritual transcendence for individuals from different cultural and religious backgrounds. However, we believed that experiences of selfless, spiritual transcendence would be interpreted differently based on inherent cultural and religious influences, similar to the manner by which verbal language skills are generally neurologically hardwired in the left hemisphere but expressed differently based on cultural influences. Specifically, the vast majority of humans (~90%) have general language skills primarily located in the left hemisphere (e.g., Broca's and Wernicke's areas); however, a person's native language is primarily determined on social and cultural contexts (i.e., if you are raised in China, you will likely speak Chinese; if you are raised in Germany, you will likely speak German).

We proposed that the same should be true for the religious expression of RPL-based selflessness. In general, humans primarily process information related to their self in the right hemisphere/parietal lobe. As such, reduced functioning of the RPL should be associated with a decreased ability to focus on oneself and to experience increased spiritual transcendence, regardless of one's cultural background or nationality. However, the manner in which selfless transcendence is experienced will differ based on individual's cultural and religious backgrounds. That is, Christians will report a strong or deep connection with Christ, Muslims an increased connection with Allah, and Hindus increased connections with any of a multitude of deities specific to their devotional emphasis (whether toward Vishnu, Shiva, Kali, etc.).

With these issues in mind, we completed a study (Johnstone, Bhushan, Hanks, Yoon, & Cohen, 2016; also see Johnstone, Hanks, et al., 2016) of the spiritual transcendence of 109 individuals with TBI from different cultures and languages (India and the United States), ethnicities (African American, Caucasian, and South Asian), and religions (Christian, Hindu, and Muslim). In the study, we included Caucasian Christians from central rural Missouri, African American Christians from urban Michigan, and South Asian Hindus and Muslims from Kanpur, India. The diversity of our sample was considerable.

We translated the BMMRS into Hindi, one of the main languages of India and used widely in northern India especially, and our colleagues in India administered this and the standard neuropsychological measures (English or Indian versions) to all participants. This cross-cultural study was limited in certain ways, including the use of a measure of spirituality designed primarily for western Christians (i.e., BMMRS) with an Indian sample, with the assumption that spiritual items had similar meanings for individuals from different Asian- and Western-oriented religious traditions. However, even with these limitations, the results provided interesting findings that nonetheless generally supported the original hypotheses.

The results (see Table 4) indicated that decreased RPL functioning (as measured by the JOLO test) was significantly and negatively correlated with four of five BMMRS spiritual scales for the entire sample (and it did approach significance for meaning). As expected, decreased RPL functioning (and inferred increased selflessness) was associated with increased closeness with the divine for individuals from multiple cultural, ethnic, religious, national, and linguistic backgrounds. In addition to the consistency of the JOLO test results, the findings were also generally consistent with our previous studies in that a measure of left temporal lobe functioning was positively (and significantly in this study) correlated with four of

TABLE 4 Correlations Between Spirituality and Neuropsychological Indexes of Cerebral Functioning

Neuropsychological	DSE	Values/belief	Meaning	Forgiveness	RS coping
RIGHT HEMISPHERE					
Right parietal lobe					
JOLO	−0.29**	−0.27*	−0.18	−0.20**	−0.21*
LEFT HEMISPHERE					
Left temporal lobe					
WMS-III LM I	0.37**	0.24*	0.17	0.35**	0.36**

*$p < 0.05$; **$p < 0.01$; *BMMRS*, Brief Multidimensional Measure of Religiousness/Spirituality; *DSE*, Daily Spiritual Experience; *RS*, Religious/Spiritual Coping; *JOLO*, Judgment of Line Orientation; *WMS-III*, Wechsler Memory Scale-III. See Johnstone, Bhushan, et al. (2016) and Johnstone, Hanks, et al. (2016) for specific information about the neuropsychological tests used and their demonstrated relationships to specific cerebral regions.

the five BMMRS spiritual scales (and approached significance with the meaning).

Although the results were as hypothesized and statistically significant for the entire group, they were not significant for each of the individual religious groups, possibly related to limited statistical power. However, consistent with our original conjectures, the overall results further suggested that decreased RPL-related selflessness operates as a universal neuropsychological foundation for spiritual transcendence across multiple cultures, ethnicities, and faith traditions, even if such experiences are interpreted differently depending on religious context and background. The implications of these findings were significant in that they suggested that the propensity for spiritually transcendent experiences, regardless of ethnicity, culture, or religion, is generally available to all individuals and based on decreased RPL functioning and increased selflessness. It was nice to have results that emphasized the general similarities between religious faiths, with selflessness operating as a core neuropsychological process of spiritual transcendence, despite inherent theological and doctrinal differences. It was also refreshing to note that these conclusions support the tenets of many religions throughout history that have stressed the importance of selflessness as a means to achieve spiritual development.

7 BRAIN TUMORS AND SPIRITUAL TRANSCENDENCE

While designing and conducting our research, we were reviewing the literature to learn about other research exploring the neuropsychology of spiritual transcendence and particularly any studies related to decreased

RPL functioning and increased selflessness. Whereas our previous studies were cross-sectional, limiting any inferences that can be drawn related to causation among specific neuroanatomical regions, together with our students, we found an interesting study conducted by a team of neuroscientists in Italy. Their study was longitudinal (the first in the neurological study of transcendence), and it involved individuals with operable brain tumors. The researchers used pre- and postsurgical evaluations to determine causal relationships between different brain regions and the spiritual orientations of their participants. It was probably the most comprehensive, thorough study of the neurology of spirituality to date.

Specifically, Urgesi and his colleagues (Urgesi, Aglioti, Skrap, & Fabbro, 2010) conducted a longitudinal study of 88 individuals who had neurosurgical removal (i.e., resection) of different types of brain tumors (e.g., high grade, low grade, recurrent gliomas, and meningiomas). Participants were categorized by the general location of their tumors, including the frontotemporal cortex (i.e., the anterior brain group) and the occipitotemporoparietal cortex (i.e., the posterior brain group), and also whether tumors were located in the left versus the right cerebral hemisphere. The participants of their study completed measures of religiosity and spirituality from 1 to 7 days before their neurosurgery and then once again about 3–7 days after their surgery to specifically assess any relationship between the surgical removal of their tumors and any changes in their levels of spiritual transcendence. The short time frame was used to ensure that changes in transcendence were due to tumor removal (i.e., neurologically related outcomes) and not to their long-term adaptation to surgery through environmentally related adjustments. The researchers administered the Self Transcendence (ST) Scale of the Temperament and Character Inventory (TCI; Cloninger, Pryzbeck, Svrakic, & Wetzel, 1994) to their participants to measure spiritual transcendence, which included questions such as the following:

- I often feel so connected to the people around me that it is like there is no separation between us.
- I sometimes feel so connected to nature that everything seems to be part of one living organism.
- Sometimes I have felt like I was part of something with no limits or boundaries in time and space.
- I often feel a strong sense of unity with all the things around me.
- I have had moments of great joy in which I suddenly had a clear, deep feeling of oneness with all that exists.

The researchers hypothesized that removal of sections of the frontal lobes would be related to *decreased* spiritual transcendence and that removal of certain parts of the temporoparietal areas would be related to *increased* spiritual transcendence, hypotheses that are also associated with our research—although at the time we were unaware of one another's research projects. Their results generally confirmed the hypotheses as follows. The posterior tumor group endorsed significantly higher spiritual transcendence than the anterior tumor group both before and after neurosurgery, and individuals with more severe tumors reported significantly higher levels of spiritual transcendence both before and after neurosurgery.

Although their study did not find differences in spiritual transcendence for any of the groups in terms of left- versus right-sided brain tumors, it nonetheless showed that individuals with posterior brain lesions reported higher spiritual transcendence than did individuals with anterior brain lesions. Consistent with our research, the dysfunction of the parietal lobes was associated with increased spiritual transcendence. The authors concluded that, compared with previous cross-sectional research, their longitudinal study was the first to show "a specific causative relationship between lesions of the left and right posterior parietal cortex and ST [spiritual transcendence] changes" (Urgesi et al., 2010, p. 312). The authors made several other noteworthy conclusions about relationships among the parietal cortex, selflessness, and transcendence. Specifically, they concluded that "extended self-referential awareness, which characterizes high ST [spiritually transcendent] individuals, is supported by a lower activity of the left and right posterior parietal cortex" (p. 316). Finding agreement with earlier research, they concluded further that "the reduction of neural activity in the temporoparietal cortex during spiritual experiences may reflect an altered sense of one's own body in space" (p. 316). Or stated differently, as we argue, it represents a change in one's sense of self and a strong experience of selflessness.

8 PARIETAL LOBE, TECHNOLOGY, AND SPIRITUAL TRANSCENDENCE

Through advances in medical technologies, new methods for studying the neurological foundations of spiritual transcendence have developed. Noninvasive electromagnetic fields can be administered to the brain to increase or decrease physiological activity in specific, targeted regions, allowing for the identification of specific neuropsychological processes, such as self-orientation or the sense of self, that are associated with specific neurological networks. These technologies have also been used to assess the neurophysiological foundations of spirituality, consistent with research that was described in Chapter 4 in which repetitive transcranial magnetic stimulation (rTMS) to the "inferior" or lower portion of the RPL was shown to be associated with an inability to distinguish between "self" and "others" (Uddin, Molnar-Szakacs, Zaidel, & Iacoboni, 2006).

In general, TMS allows for "temporary lesion analysis" as if the brain has an injury to a definite region, and in this way, specific neuropsychological functions can be temporarily disrupted or enhanced (i.e., either inhibited or excited) to determine related neuroanatomical-neuropsychological associations, without any permanent injury to the brain.

One of the first studies to use electromagnetic brain stimulation to induce spiritual experiences was performed by a group of Italian researchers investigating relationships among frontoparietal networks and responses to religious-spiritual stimuli (Crescentini, Aglioti, Fabbro, & Urgesi, 2014). These investigators used various forms of magnetic stimulation to either excite or inhibit the functioning of different cerebral areas and then administered different tests to determine if the excitation or inhibition affected an individual's physiological responses to religious or spiritual concepts. But rather than asking participants to self-report their subjective spiritual experiences, the researchers administered what is known as an implicit association test (IAT) focused on religiousness and spirituality. IATs are hypothesized to assess responses to stimuli before they have been reflected on consciously, as opposed to self-reported survey data, which could be biased by reflection before initiating a response. Thus, IATs purportedly assess "implicit attitudes," described as evaluative processes that occur without conscious awareness and help remove potential biases in responses. Basically, your initial physiological response is measured very quickly before you have had any time to think over your answer.

Using TMS, Crescentini and his colleagues created "virtual lesions" in the IPL and in the DLPFC of eight healthy individuals. When neurological function was inhibited in these specific regions of the brain, each participant was then administered an IAT designed to assess their (implicit) physiological responses to the religious and spiritual stimuli presented. The results indicated that inhibition of the IPL was associated with an increase in participant's implicit response to religious-spiritual stimuli, suggesting a causative role of an inhibited IPL, and in our understanding temporarily creating a more selfless individual, which then facilitated the enhanced processing of religious-spiritual stimuli and experiences.

This Italian research group then conducted a follow-up study to investigate more directly the specific role of the *right* inferior parietal lobe (RIPL) in the processing of religious-spiritual stimuli (Crescentini, Di Bucchianico, Fabbro, & Urgesi, 2015). In this study, 14 participants were given theta burst stimulation (TBS) to the RIPL of the brain while exposed to four different categories conceptualized as follows: self (i.e., I, me, or mine), other (i.e., you, yours, or his/hers), religious/spiritual (i.e., believer, God, or divine), or nonreligious/spiritual (i.e., atheist, body, or natural). The TBSs administered to the RIPL were either excitatory, inhibitory, or a placebo (i.e., a sham stimulation), and participants had to then respond (again implicitly without thinking anything over) as quickly as possible to the specific categories (e.g., self and religious/spiritual) under these different TBS conditions. As hypothesized and reaffirming their results of their first study, the researchers found that excitatory stimulation of the RIPL was associated with a significantly decreased response time for religious and spiritual words. Simply stated, the more excited and active the RIPL became, the slower the participants' implicit response to religious and spiritual words, which in our view suggests the excited RIPL led to lessened selflessness and thereby decreased religious and spiritual connection.

Taken together, these Italian studies generally demonstrate that the IPLs and particularly the RIPL are related to how we generate religious and spiritual experiences. Although not directly related to spiritually transcendent experiences, their results nonetheless provide additional support for role of the RIPL in the experience of selflessness and spiritual transcendence.

9 PULLING IT ALL TOGETHER

We find that many empirical studies of the neurological foundations of spirituality are generally supportive of our view that RPL-related selflessness is one of the neuropsychological foundations of spiritual transcendence. Overall, research has indicated a relationship between the RPL and spiritual transcendence, which has been identified using multiple measures of spirituality (e.g., Inspirit, BMMRS, Spiritual Transcendence Scale, and IATs) and multiple methods of assessing RPL functioning (e.g., JOLO and left hand finger agnosia). In addition, studies of different neurological disorders including TBI, seizures, tumors, and temporarily induced lesions have utilized a variety of brain imaging and electromagnetic technologies (e.g., SPECT, PET, MRI, fMRI, TMS, and TBS). Individuals engaging different religious practices (e.g., meditation and prayer) who come from diverse ethnic, (e.g., African Americans, Caucasians, and South Asians), cultural (e.g., the United States and India), and religious traditions (e.g., Buddhist, Christian, Hindu, and Muslim) have participated. Although simplistic, certain relationships appear repeatedly and are relatively consistent. Based on these results, our model of spiritual transcendence and selflessness was taking form as follows:

- Individuals experience a "sense of self" that is the result of the integration of neuropsychological processes associated primarily with right-hemisphere and especially parietal functions.
- This sense of self may be minimized, either through injury, disease, or behavioral practices, leading to heightened experiences of selflessness.
- These experiences of selflessness may be interpreted as occurrences of spiritual transcendence, depending on context and background.

Although these empirical and neurological studies tended to support our model, we believed it was necessary to next determine how the potent transcendent experiences of religious mystics and even of those who are not necessarily religiously oriented are expressed and what their insights might tell us about the relationship between selflessness and profound spiritual experience. Our hope was that these collective insights would fit with our model of the connection between selflessness and spiritual transcendence and help make it stronger by generating a productive conversation between neuroscience and the humanities to enhance our understanding about the nature of spiritual transcendence.

SELFLESSNESS AS THE KEY TO TRANSCENDENCE

Faith Traditions, Spiritual Transcendence, and Selflessness

Theologians may quarrel, but the mystics of the world speak the same language ...
(Attributed to Meister Eckhart (German theologian, 1260–1328), actually from Easwaran, 1996/1989, p. 24)

1 RELIGION, SPIRITUALITY, AND MYSTICAL EXPERIENCE

Many religions teach that the human self is limited and that there are ways to overcome that limited self through transcendent experiences. This theme is found in many religions especially in the context of profound religious experiences that may be foundational to the theological framework of a faith tradition. In a statement attributed to the medieval founder of the Franciscans, St. Francis of Assisi (1181–1226), he indicates

109
© 2019 Elsevier Inc. All rights reserved.

that, "Above all the grace and the gifts that Christ gives to his beloved is that of overcoming self." One question to ask is whether there is some form of common ground or connection that may help account for this type of potent religious and spiritual experience. Can the insights of diverse faith traditions be helpful in creating a fuller and more comprehensive understanding of spiritual transcendence that could inform the sciences in exploring the nature of these experiences? We believe that exploring experiences of selflessness from the perspective of different religions will shed new light on neuroscientific and neuropsychological research of spiritual transcendence. By integrating research from religious perspectives on transcendence with scientific insights into what happens in the brain during experiences of transcendence, we hope to create a deeper understanding of an essential part of our humanity, generating positive interactions between the humanities, the social sciences, and the natural sciences to facilitate a higher level of interdisciplinary communication and integrative approaches to research.

It appears as more than coincidence that many mystics and luminaries from different faith traditions have implicated the loss of self in attaining the highest insights and spiritual rewards of their respective religious orientations. Sometimes, these mystical experiences of the loss of self have even laid the foundations for entirely new religious outlooks or faith traditions. The experiences of individuals who acknowledge occurrences of spiritual transcendence often connect their profound loss of sense of the self with making significant strides in their spiritual development. These outcomes may even influence the broader theological and philosophical outlooks that develop within different religious traditions. Our contention is that the relationship between the loss of self and spiritual transcendence can best be understood from an interdisciplinary perspective, integrating insights from both the humanities and science.

The previous chapters build the argument that the brain processes a sense of the self, primarily in the right hemisphere, which can be increased or decreased depending on injury or behavioral practices. We contend that these selfless neuropsychological processes are universal for all individuals across all religions but that they are experienced differently depending on religious upbringing. In this chapter, we provide examples of mystical experiences reported by individuals from major faith traditions and religious texts that indicate the role of self/selflessness in experiencing transcendence and achieving ultimate salvation.

Faith traditions often promote experiences of spiritual selflessness as an exemplary human trait and as an experiential necessity for deepening consciousness, achieving higher states of spiritual awareness, and ultimately for attaining salvation. Individuals from various religions throughout history have described their spiritually transcendent experiences in different ways. They may have a sense of oneness or union with God or perhaps a deep sense of connectedness to nature or the universe, but these experiences often involve a significant reduction or even complete loss of aspects of the personal and immediate sense of the self. Our model suggests that these types of experiences of spiritual transcendence are in part related to the reduction of the specific neuropsychological processes of self-orientation associated with the right hemisphere and the right parietal lobe. Experiences of spiritual selflessness, when viewed collectively from neuropsychological *and* religious perspectives, provide a powerful bridge that connects the sciences and the humanities, and through this exploration, the humanities can help drive neuroscientific research to facilitate a better understanding of human experiences of profound spiritual transcendence.

2 SELFLESSNESS IN RELIGIOUS TRADITIONS: THE BUDDHIST "NON-SELF"

This is not mine, I am not this, this is not my self.
Pali Canon, Sayings of the Buddha *(Upadaparitassana Sutta, Samyutta Nikaya 22.8)*

We first use the example of Buddhism, in part because Buddhism teaches that the self is nothing but a problematic illusion. Before the Buddha attained enlightenment, he was a Hindu prince and heir to his father's kingdom. But at the age of 29, he abruptly left the palace, leaving behind his wife and infant son, became an ascetic and a mystic, and wandered in the wilderness seeking higher knowledge and the truth of existence. At the age of 35, after 6 years of arduous ascetic lifestyle, the Buddha found that the truth he was seeking was rooted in overcoming the false sense of the self. With his realization of ultimate selflessness, he attained enlightenment and then spent the rest of his life sharing this truth (*dharma*) with others, so they too could strive for salvation.

Buddhism understands the sensory, cognitive, and consciously oriented sense of self as a misunderstanding of the inherently false reality of the "self," which according to Buddhist understanding is in truth empty (*shunya*) and only exists as a product of dependent arising (*pratitya samutpada*) of the perceived "self" explained through the composition of five hierarchical aggregates (*skanda*), with the highest of these being consciousness. These aggregates are not things but rather processes in constant flux. In the Buddhist perspective, there is no essential or real self, because the self is simply a mental construction and is constantly changing, such that what might even seem to be the self at one moment immediately becomes something different in the next. Due to desires, attachments, and mental impurities, humans project a false sense of permanent self that hinders them from understanding the true nature of reality, including the implications of reincarnation and the path to salvation. The attachment to this illusory self is a serious obstacle on the path to salvation. But this misconception can be overcome and as the Buddha is quoted as saying, in the context of knowledge and insight that could be gained through the practice of meditation:

> The perception of impermanence, when developed and increased ... wears down and destroys the conceit of "I am." *Pali Canon, Samyutta Nikaya* III.155–6 *(also quoted in Collins, 1982, p. 114)*

Thus, by continually developing through spiritual practices, an increasing awareness of true selflessness and the truth of the transitory nature of the "self" can be discovered. Gradually, a realization develops that there is no essential self (framed in the Buddhist concept of *anatman*, which in Sanskrit literally means "no self") and that the conditioned experience of selfness is not an accurate reflection of the true nature of one's existence. As Master Dogen (1200–53 CE), the founder of the Soto school of Zen, said,

> To study the way of the Buddha is to study your own self. To study your own self is to forget yourself. To forget yourself is to have the objective world prevail in you. Master Dogen *(cited in Austin, 2009, p. 49)*

With the full attainment of this insight, namely, that self lacks any real essence, a unique awareness arises, and eventually, enlightenment (*bodhi*) and *nirvana* become possible as the result of this spiritual awakening.

What is interesting about Dogen's statement, quoted above, is the point that as you study your "self," you learn to forget or overcome the (false sense of) self. In Mahayana and Vajrayana Buddhism, the false sense of the self has been explicitly extended beyond the self to all things, such that no "things" have true essential qualities or real essence and that they too are in truth "empty" (*shunya*), just as is the self. This is consistent with both the Buddhist and physical sciences' conceptualization of impermanence (*anitya*), meaning that everything, including the self, is always in a constant state of change and therefore has no absolute sense that always holds steady and defines their essence.

This dismantling of the self and the complete merging of person with the constantly changing objective world are similar in some ways to what occurs in other religious traditions with certain experiences of spiritual transcendence, as the self becomes obliterated in part by connecting to something larger than itself. Buddhism is (often) atheistic, and the practitioner does not merge with higher powers, with divine entities, or with God. Even modern physics and cosmology would seem to agree with Buddhism that the universe is in constant flux and that, for example, all matter in the universe has an impact on all other matter, even if that influence is miniscule and immeasurable. This is evident in the following discussion between the Buddha and his disciple Ananda:

> "It is said that the world is empty, the world is empty, lord. In what respect is it said that the world is empty?" The Buddha replied, "Insofar as it is empty of a self or of anything pertaining to a self: Thus it is said, Ananda, that the world is empty." *Pali Canon, Samyutta Nikaya, 35.18 (Bhikkhu, 1997)*

This conception of emptiness and selflessness is not, in contrast to depersonalization, a fundamental aversion to experience and existence. Rather, it is a transcendent awareness that can be awe-inspiring or even mystical, and in the Buddhist tradition can lead a person toward ultimate salvation or *nirvana*. The teachings of Buddhism are clear in this regard (see, e.g., Collins (1982) on "selfless persons"). As the Buddhist monk Joshu Sasaki Roshi has noted,

> There is nothing true about Buddhism, except for the manifestation of no-Self. Joshu Sasaki Roshi *(quoted in Austin, 2009, p. 85)*

All traditions of Buddhism agree that any assertion of an essential self is illusionary and in truth empty, and misunderstanding in this regard is the strongest basis of all suffering (*duhkha*) in the world. Suffering is caused by a failure to comprehend or realize the truth of the transitory and impermanent nature of the self (and the universe) expressed through the Buddhist concept of impermanence. Thus, selflessness in Buddhism is attained progressively as one overcomes the falsely conditioned sense of self and gradually eliminates suffering (*duhkha*) caused by conditioned attachments to the world. Advancement is made by following the "Eight-Fold Path" that the Buddha codified in the Four Noble Truths and elaborated to his followers.

Understanding of the false sense of self and increased selflessness is developed and cultivated (*bhavana*) by a Buddhist practitioner's long-term efforts engaging in different practices of meditation (*dhyana*). These well-established ritual procedures help develop and enhance the ability both to hone mental concentration (*samadhi*) and to gain true insight (*vipasanna*) through the achievement of higher levels of mindfulness (*sati*) and thereby help secure real understanding and wisdom (*prajna*) needed to eventually overcome the false sense of self.

3 THE HINDU SELF AS TOTALITY

> As soon as I think that I am a little body, I want to preserve it, to protect it, to keep it nice at the expense of other bodies; then you and I become separate. First get rid of the delusion "I am the body," then only will we want real knowledge.
> *Swami Vivekananda (1863–1902), The Complete Works of Swami Vivekananda (vol. 2)*

III. SELFLESSNESS AS THE KEY TO TRANSCENDENCE

> When he perceives the unity
> existing in separate creatures
> and how they expand from unity,
> he attains the infinite spirit (*Brahman*).
> *Bhagavad Gita XIII, 30 (Miller, 1986)*

In contrast to Buddhism, in Hinduism, there is a strong recognition of an eternal "self" (*atman*), which is regarded as constituting a person's true essence. The Hindu self is regarded as something that is reincarnated through a great multitude of lives and existences. However, in divergence with Buddhism, which also holds a belief in reincarnation, in Hinduism, there is said to be an eternal self (*atman*) that is reincarnated repeatedly from one lifetime to the next. According to Hinduism, this eternal core self is never lost, nor can it in any way be changed, manipulated, or destroyed. As the famous Hindu devotional text the *Bhagavad Gita* says,

> Our bodies are known to end,
> but the embodied self (*atman*) is enduring,
> indestructible, and immeasurable ....
> *Bhagavad Gita II, 18 (Miller, 1986)*

> As a man discards
> worn-out clothes
> to put on new
> and different ones,
> so the embodied self (*atman*)
> discards
> its worn-out bodies
> to take on other new ones.
> *Bhagavad Gita II, 22 (Miller, 1986)*

While Hinduism and Buddhism have radically different notions of the self, they do nevertheless have similar ideas about transcendence in terms of eliminating all *karma* (i.e., the residues of one's actions). Additionally, the ultimate goal of both religions is to escape from the continuing life/death experiences of reincarnation. However, in contrast to Buddhism, Hinduism believes in an eternal "self" (*atman*), which constitutes a person's true essence, and that it is this self that is reincarnated repeatedly through successive lifetimes.

The ancient sacred Hindu texts called the *Upanishads* propose that the path to salvation lies in a correct understanding of the self (*atman*). The self, according to these texts, is not identical to the narrow human "ego," but rather to the cosmic universal totality (*Brahman*). To attain salvation, a person has to recognize that the *atman* and *Brahman* are one. This important conception of the self became promulgated in the famous and well-known *Advaita* (literally, "nondual") *Vedanta* philosophy espoused by one of the most famous and foremost philosophers and reformers of

Hinduism, Adi Shankara, who lived in the 8th century. The defining characteristic of his nondual philosophy is that there is a metaphysical equivalence between the self (*atman*) and the cosmic universal totality (*Brahman*). In other words, the self and the universe are isomorphic, and to attain salvation, one has to spiritually recognize this interconnection.

Hindu sages and seekers who renounced the hustle bustle of the everyday world lived simple and often reclusive lives, searching for ultimate truth. Over time, ritual traditions were developed for passing this sacred (and secret) knowledge on orally to followers, through the generations, from teachers (*guru*) to their disciples through oral recitation that became the *Upanishads*. As this process continued and become more elaborated, different religious rituals and devotional practices developed, as disciples sought to grasp the Hindu theological understandings of the eternal self.

A famous *Upanishad*, the *Mandukya*, discusses four states of the self: waking consciousness, sleep with dreams, dreamless sleep, and a mystical state simply called *turiya* (or "the fourth"). Intriguingly, modern neuroscience has identified three states of consciousness that correspond exactly to the first three states of self-awareness identified in the *Mandukya Upanishad*. The fourth state described is one of transcendental consciousness and complete awareness in which all divisions and dualities disappear and the self alone exists in its pure state. As the *Mandukya Upanishad* says,

> They consider the fourth quarter as perceiving neither what is inside nor what is outside, nor even both together; not as a mass of perception, neither as perceiving nor as not perceiving; as unseen; as beyond the reach of ordinary transaction; as ungraspable; as without distinguishing marks; as unthinkable; as indescribable; as one whose essence is the perception of itself alone; as the cessation of the visible world; as tranquil; as auspicious; as without a second. That is the self (*atman*), and it is that which should be perceived. **Mandukya Upanishad 7 (Olivelle, 1998)**

Importantly, in *Advaita Vedanta* philosophy, the self or *atman* is seen as being completely identical with *Brahman*, and it is simply the failure to spiritually recognize this nonduality that creates ignorance (*avidya*) and prevents ultimate salvation (*moksha*) of the self (*atman*) by releasing it from the continual cycle of reincarnation (*samsara*). Achieving this requires a complete breakdown of all separation between the self and other.

Although transcendence in Hinduism is equated with a profound realization of self, it is nevertheless possible to characterize this state as "selfless" in one sense, since there is a loss of the lower self of the "ego." Here, we may refer to the absolute self as "the *Atman* with capital A" and the personal sense of self as "the *atman* with a small a." The *Isha Upanishad* makes this distinction between the little self (*atman*) and the transcendent self (*paramatman*) clear. As the *Isha Upanishad* indicates,

III. SELFLESSNESS AS THE KEY TO TRANSCENDENCE

> When a man sees all beings
> within his very self (*atman*),
> and his self within all beings,
> it will not seek to hide from him.
> When in the self of a discerning man,
> his very self has become all beings,
> what bewilderment, what sorrow can there be,
> regarding that self of him who sees this oneness.
> *Isha Upanishad 6–7 (Olivelle, 1998)*

Thus, the subjective experiences of the personal self (*atman*) are merely ephemeral fluctuations in the underlying unchanging essence of the Self (*Atman*). As all superficial senses of the phenomenal/incarnate self are overcome, the unification of manifest and the unmanifest is attained with the realization that *atman* equals *Brahman*. The *Mundaka Upanishad* explains

> … Works and the self consisting of knowledge—
> all unite in the highest immutable.
> *Mundaka Upanishad III:2,7 (Olivelle, 1998)*

Just as self-identity disappears as one connects with totality, a river loses its individuality as it flows into the ocean:

> As the rivers flow on and enter into the ocean
> giving up their names and appearances;
> So the knower, freed from name and appearance,
> reaches heavenly Person, beyond the very highest.
> *Mundaka Upanishad III: 2,8 (Olivelle, 1998)*

Complete selflessness and a higher spiritual awareness are achieved as one fully unites with the impersonal cosmic principle *Brahman*:

> When a man comes to know that highest *Brahman*, he himself becomes that very *Brahman*.
> *Mundaka Upanishad III:2,9 (Olivelle, 1998)*

This highest spiritual state, in other words, involves the complete loss of personal self or total selflessness where there is no longer any separation between personal self and cosmos/totality (*Brahman*). It is a process that is said to take innumerable lifetimes to accomplish, but once achieved, the perceived separation between subject and object and self and nonself is broken down completely, and ultimate salvation (*moksha*) of the eternal self (*Atman*) becomes possible. When this liberation occurs, the self no longer reincarnates, and its existence in the cycle of reincarnation (*samsara*) is ended as it escapes from any future reincarnation. It is often described as a state of total peace, complete liberation, and final release.

It is impossible to engage here in a more detailed discussion of the complex conceptual developments that have occurred philosophically in the

Hindu (and Indian) idea of the "self" over time (see e.g., Gardner, 1998; Siderits, Thompson & Zahavi, 2011).

4 CHRISTIANITY, SOUL, SPIRIT, AND SELF

Above all the grace and the gifts that Christ gives to his beloved is that of over-coming self.
Saint Francis of Assisi (1181–1226; founder of the Franciscans)

What is the human "self" according to Christianity? Notions of "self" have changed in Christianity over time and can be difficult to identify with precision. The New Testament, which was originally written in Greek, uses two different words for the human self: "soul" (*psyche*) and "spirit" (*pneuma*), with these both encased in a physical body (*soma*). For example, "Now may the God of peace Himself sanctify you completely, and may your whole spirit (*pneuma*), soul (*psyche*), and body (*soma*) be preserved" (1 *Thessalonians* 5:23).

The word "spirit" is often used in the sense of "life" (e.g., *Matthew* 27:50, "And Jesus cried out again with a loud voice and yielded up His spirit"; *Luke* 8:55, "Then her spirit returned, and she arose immediately") but can also refer more directly to the disposition of the individual self (e.g., *Matthew* 26:41, "the spirit indeed is willing, but the flesh is weak"). As the apostle Paul says, it is the spirit that is capable of "knowing" things, "For what man knows the things of a man except the spirit of the man which is in him?" (1 *Corinthians* 2:11).

In the New Testament, selflessness is often connected to the notion of merging one's spirit with the divine Spirit of God. In 2 *Corinthians*, Paul suggests that the union with the Spirit of God leads to transformation; Paul says

Nevertheless when one turns to the Lord, the veil is taken away.
Now the Lord is the Spirit, and where the Spirit of the Lord is, there is liberty.
But we all, with unveiled face, beholding the glory of the Lord, are being trans-formed into the same image from glory to glory, just as by the Spirit of the Lord.
2 Corinthians 3:16–18, New King James Version

Here is a classic example from the New Testament, where union with God occurs when one's spirit is completely immersed in the divine spirit:

But he who is joined to the Lord becomes one spirit with him.
1 Corinthians 6:17, New King James Version

One's regular awareness of the spirit is lost by merging it into some-thing profoundly larger, as one's spirit is merged together with the Spirit of God.

III. SELFLESSNESS AS THE KEY TO TRANSCENDENCE

In a well-known biblical story, Paul had an intense conversion experience on the road to Damascus. He was suddenly surrounded by light, fell down, and heard a voice speaking to him:

> As he journeyed, he came near Damascus, and suddenly a light shone around him from heaven. Then he fell to the ground, and heard a voice saying to him, "Saul, Saul, why are you persecuting Me?"
> And he said, "Who are You, Lord?"
> Then the Lord said, "I am Jesus, whom you are persecuting. It is hard for you to kick against the goads."
> *Acts 9: 3–5, New King James Version*

Paul does not appear to lose his sense of self as he becomes surrounded by some sort of unique light (different from daylight); he falls to the ground and then hears a voice that he apparently knows is divine. Then toward the end of this strange incident, Paul becomes blind temporarily and seems to be in some sort of trance state where he must be physically helped by others. Some have suggested that this episode might be evidence of an epileptic seizure, and there are cases of religious conversion thought to be connected to epilepsy (Dewhurst & Beard, 2003; Devinsky & Lai, 2008; Saver & Rabin, 1997). Paul's sense of himself or who he is abruptly and forever changed by this event, and his sensation of this event's reality is unquestioned—strong evidence of a profound spiritual and mystical experience.

In the New Testament and in other writings of early Christianity, experiences of spiritual transcendence are sometimes expressed as coming under the influence of a special kind of light or luminosity (as was seen in the case of Paul). Saint Augustine of Hippo (354–430 CE) wrote in his famous *Confessions*:

> … I entered even into my inward self, Thou (i.e., God) being my Guide: and able I was, for Thou wert become my Helper. And I entered and beheld with the eye of my soul (such as it was), above the same eye of my soul, above my mind, the Light Unchangeable. Not this ordinary light, which all flesh may look upon, nor as it were a greater of the same kind, as though the brightness of this should be manifold brighter, and with its greatness take up all space. Not such was this light, but other, yea, far other from these. Nor was it above my soul, as oil is above water, nor yet as heaven above earth: but above to my soul, because It made me; and I below It, because I was made by It. He that knows the Truth, knows what that Light is; and he that knows It, knows eternity.
> *(The Confessions of Saint Augustine, book VII, 401 AD; E. B. Pusey, translation)*

Here, we see not only an example of losing oneself by connecting with God and being completely encompassed but also a spiritually transcendent experience that connects one with "eternity."

As Christianity develops monastic traditions, it fosters ritual traditions of prayer and meditation for the unification of the individual with

God through intense experiences of spiritual transcendence. The ground-work for this development occurs in early Christianity among the Desert Fathers, early Christian hermits and ascetics who lived mainly in the desert of Egypt beginning around the third century CE. Over time, thousands of monks and nuns were drawn to these Christian desert communities in the quest for spiritual salvation. These communities became the precursor for the development of the contemplative practices of medieval monastic Christianity often used to foster profound experiences of spiritual transcendence.

During the 14th century, *The Cloud of Unknowing*, an anonymous text on Christian mysticism, appears and quickly becomes very popular. This text is a guide on contemplative prayer as a spiritual practice written during the height of European monasticism. Its target audience is young monastics, and it is an instructional text (not a didactic work) that calls on the reader to grow closer to God through specific rituals of meditation and prayer that will generate profound mystical experiences. According to the text, one has to be courageous enough to surrender one's mind and ego to the "cloud of unknowing" in order to glimpse the nature of God. In Chapter 44, *The Cloud of Unknowing* asks, "How a soul shall dispose it on its own part, for to destroy all witting and feeling of its own being?" In other words, how can one lose one's immediate sense of self and connect with God. The advice given is as follows:

> BUT now thou askest me, how thou mayest destroy this naked witting and feeling of thine own being. For peradventure thou thinkest that an it were destroyed, all other lettings were destroyed: and if thou thinkest thus, thou thinkest right truly. But to this I answer thee and I say, that without a full special grace full freely given of God, and thereto a full according ableness to receive this grace on thy part, this naked witting and feeling of thy being may on nowise be destroyed.
> *The Cloud of Unknowing (Underhill, 1911, p. 211)*

It indicates that one must be able to lose one's immediate sense of self (wit and feeling) if one wants to receive the special grace of God.

Just prior to the popularization of *The Cloud of Unknowing*, the German theologian Meister Eckhart (*c*.1260–1328), a member of the Dominican Order, wrote about the need to go beyond time and space in order to know God:

> Nothing hinders the soul's knowledge of God as much as time and space, for time and space are fragments, whereas God is one. And therefore if the soul is to know God, it must know him above time and outside space; for God is neither this nor that as are all those manifested things.
> *Meister Eckhart (quoted in Stace, 1960a, p. 196)*

As we have argued in earlier chapters, the self can be conceptualized in terms of time and space (i.e., the self is defined as a specific body that

III. SELFLESSNESS AS THE KEY TO TRANSCENDENCE

occupies a specific physical space at a specific time). If "the self" is substituted for "time and space," in Eckhart's quote, it now reads

> Nothing hinders the soul's knowledge of God as much as "the self," for "the self" is/are fragments, whereas God is one. And therefore, if the soul is to know God, it must know him above "the self"; for God is neither this nor that as are all those manifested things.

Yet, Eckhart is aware that during the loss of self in powerful transcendent experiences, there must also be a way to be able to recollect the occurrence of such profound spiritual experiences. As Eckhart says,

> In this exalted state she (i.e., the soul) has lost her proper self and is flowing full-flood into the unity of the divine nature. But what, you may ask, is the fate of this lost soul? Does she find herself or not? ... It seems to me that ... though she sink in all the oneness of divinity she never touches bottom. God has left her one little point from which to get back to herself ... and know herself as creature.
> *Meister Eckhart (quoted in Stace, 1960a, p. 114)*

Eckhart came under fire from the Catholic authorities for the radical ideas he presented about mystical union and is said to have recanted any offenses to the Church right before his death. Nonetheless, his emphasis on losing the self, by merging one's soul more strongly with God in order to find God's presence, had taken hold.

Building on this upsurge of Catholic mystical contemplation, the Carmelite monastic communities of Saint Teresa of Avila (1515–82) and Saint John of the Cross (1542–91) provide good examples of the development of a mystical contemplative Christianity involving deep prayer and surrendering oneself to God. Saint Teresa was a Carmelite nun known for her intense bouts of mystical absorption. In describing her union with God, she wrote as follows:

> It is plain enough what union is—two distinct things becoming one.
> *Saint Teresa (quoted in Stace, 1960a, p. 221)*

Saint John said that

> The soul must be emptied of all these imagined forms, figures, and images, and it must remain in darkness in respect to these ... the alienation and withdrawal of the spirit from all things, forms, and figures, and from the memory of them.
> *Saint John (quoted in Stace, 1960a, p. 103)*

As this process occurs, Saint John indicated that

> The more the soul learns to abide in the spiritual, the more comes to a halt the operation of the facilities in particular acts, since the soul becomes more and more collected in one undivided and pure act.
> *Saint John (quoted in Stace, 1960a, p. 103)*

But this mystical upsurge in monastic Christianity was at times (and still today) often seen as a threat to ecclesiastical authority because it challenged the neatness of reason-based theology and the orderly operation of the institutional church (John-Julian, 2015).

Notably, not only was the nature of the soul's interactive relationship with God showing changes but also the concept of the person found in early Christianity and during the Middle Ages. These conceptions would change even more profoundly as new ideas about the self and consciousness developed in the philosophy of the Reformation period and in the Age of Reason (Winquist, 1998). This transition cannot be described in detail here except to say that a revolution occurred as the authority underlying Christian spirituality shifted significantly from institutional interpretation to individual experience, from which much of the modern conception of the Christian self has emerged. The self becomes increasingly subjectified in the developing Christian experience of spiritual self-awareness. This lays the foundation for the development of the modern notion of the self, as the self now becomes something that is known subjectively and experienced directly via self-reflection.

In the post-Reformation era, with the emergence of Protestant forms of Christianity, expressions of Christian religious faith continued to move from formalized outward acts and rituals to a more inward disposition where religious values increasingly become a personal issue and belief a fundamentally interior state with a psychological basis (Ruel, 1982). This inward shift signified an important movement leading into the Enlightenment philosophies of consciousness where self-transcendence is equated with immanence and selflessness and experiences of connectedness to the whole or totality of nature and the universe. In Christian spirituality, this movement is reflected in a changing sense of the nature of salvation. Concerns tied to salvation initially centered on original sin and the first, yet fallible, humans Adam and Eve created in the image of God and later banished from paradise; later, it shifts more onto the individual whose salvation can only be restored through direct communion with Jesus, the divine in human form. To achieve salvation, a Christian must become selfless, giving oneself wholly to Jesus without reservation—but in a way that is articulated differently from what is seen in early Christianity.

Catholicism was also not immune from these reorientations of the self. As the American Catholic Trappist monk Father Thomas Merton (1915–68) has said,

> To say that I am made in the image of God is to say that Love is the reason for my existence, for God is love. Love is my true identity. Selflessness is my true self. Love is my true character. Love is my name.
>
> *Merton (1949, p. 60)*

III. SELFLESSNESS AS THE KEY TO TRANSCENDENCE

Here, selflessness represents divine, unconditional, volitional, and self-sacrificing love (*agape*). The core point is that in Christianity, one's self-abandonment or complete loss of the ego-based self to God (i.e., complete selflessness) becomes critical to achieving salvation. Merton's understanding of selflessness was developed by a powerful revelation he experienced one day while walking in a downtrodden area of downtown Louisville, Kentucky. He described his life-inspiring revelation as follows:

> At the center of our being is a point of nothingness which is untouched by sin and by illusion, a point of pure truth, a point or spark which belongs entirely to God, which is never at our disposal, from which God disposes of our lives, which is inaccessible to the fantasies of our own mind or the brutalities of our own will. This little point of nothingness and of absolute poverty is the pure glory of God in us ….
> *Merton (1965, p. 155)*

Merton understood the critical importance of having transcendent experiences that involved losing oneself in order to find one's deeper spiritual nature:

> Art enables us to find ourselves and lose ourselves at the same time. The mind that responds to the intellectual and spiritual values that lie hidden in a poem, a painting, or a piece of music, discovers a spiritual vitality that lifts it above itself, takes it out of itself, and makes it present to itself on a level of being that it did not know it could ever achieve.
> *Merton (1983/1955, p. 35)*

Recently, another Trappist, Father Thomas Keating, has revived the "Centering Prayer" movement that was popular among some of the late medieval monastic communities in contemporary Christianity, although not without some strong disapproval from the Church. Tensions between mystical endeavors and mainstream Christianity continue, as Cardinal John Henry Newman (1801–90), a well-regarded Anglican priest, poet, and theologian, stated "Mysticism begins in mist and ends in schism" (cited in John-Julian, 2015, p. xv). There appears to be an inbuilt fear of mystical orientations in the Catholic mainstream, lest these mystical (spiritually transcendent) practices become uncontrollable and lead to divisions in the Church. However, even Pope Francis has recognized the importance of mystical experience in religion and how profound its effects can be on a person. He recently said

> A religion without mystics is a philosophy …. The mystic manages to strip himself of action, of facts, objectives and even the pastoral mission and rises until he reaches communion with Beatitude. Brief moments but which fill an entire life.
> *Pope Francis (cited in John-Julian, 2015, p. xv).*

As Karl Rahner, the renowned German Jesuit theologian, wrote in the middle of the 20th century,

> It must be made intelligible to people that they have an implicit but true knowledge of God—perhaps not reflected upon and not verbalized; or better expressed, they have a genuine experience of God ultimately rooted in their spiritual existence, in their transcendentality, in their personality, or whatever you want to call it.
> *Karl Rahner (cited in Egan, 2013, p. 45)*

Rahner understood that the movement of modern religious focus on the self vis-a-vis salvation was becoming increasingly a self-reflective process with spiritual proclivities that express universal human potentials. He also saw that transcendent spiritual experience is a requirement for the survival of Christianity and perhaps humanity. As Rahner notes,

> The devout Christian of the future will be a "mystic," one who has "experienced" something, or he will cease to be anything at all.
> *Karl Rahner (cited in Egan, 2013, p. 51)*

5 JUDAISM, THE SOUL, AND CONTACT WITH GOD

> A person who truly attains selflessness emanates an awareness of the Almighty to all that surround him and come in contact with him. It is almost as if he or she has become a placeholder, a symbolic pointer, to the omnipresent holiness of the Almighty.
> *Ginsberg (2005)*

Jewish mysticism has developed outside of mainstream Judaism. Known as *Kabbalah*, literally meaning "parallel/corresponding" or "received tradition," it is a form of Jewish mystical thought based on a text called the *Zohar*. The *Zohar*, written mostly in Aramaic, is a mystical commentary on the Hebrew Torah that first appeared in 13th century Spain but is said to date back to the second century. An esoteric variant of mystical Judaism that developed in 18th century Europe, influenced by French and Spanish Jewish mysticism of the 12th and 13th centuries, is popular especially among certain groups of Hasidic Jews.

In developing its own mystical versions of Kabbalah Judaism, the Chabad Hasidics claim that each person has two different aspects of the soul (in Hebrew, *nephesh*). There is the divine part of the soul and the animal part of the soul. The individual self that we perceive in our regular conscious awareness is somewhere in between these two qualities of the soul. This means that neither the divine nor the animal parts of the soul fully define the self, and this middle part of the individual, called "*beinoni*," is literally the intermediate self. This intermediate subjective self is said to be in a constant struggle between the other two disparate aspects of the soul, one divine and the other animal-based. The yearning for authentic goodness (and to experience divine consciousness) comes from the divine soul, and opposed to this are evil inclinations empowered by the animal soul.

Extensive discussion of this topic is found in the *Tanya*, written by Rabbi Shneur Zalman of Liadi, the founder of Chabad Hasidism, and published in 1797. It defines Hasidic mystical psychology and theology and is used as a handbook for daily spiritual life in much of Hasidic Jewish observance. These writings indicate that the feeling of self that accompanies one's good actions craves acknowledgment for these actions and is a quality of the sense of self that originates in the animal soul. The intellect will also try to defend these feelings coming from the animal soul by arguing that it is natural to feel a sense of self. However, the more one "feeds" this false protected sense of self, the more one's cravings (whether permitted or forbidden) will be strengthened and intensified. The intellect will continue to make these problematic arguments until one becomes able to reorient their intellectual faculties in line with the divine consciousness of their divine soul. This feat can only be achieved by overcoming the false sense of self and realizing one's righteous qualities through ritual and behavioral forms of self-abnegation. This is where the cultivation of selflessness becomes crucial in this form of Judaism.

In this Jewish tradition, to fully realize the self (in Hebrew, *nephesh*, literally "self," "life," "soul," "person," etc.) requires the cultivation of selflessness. In this regard, there are three ideal aspects of the self that constitute Judaism's response to the demands of being ethical (Ducoff, 1989). First, there is the norm of self-realization coupled to living virtuously and helping others. Second, there is the self of mutual sharing and loving that enables each person to mutually give and receive. Third, there is the altruistic self-reflecting sacrifice, which enables the individual to truly realize the richness of both the self and selflessness.

In Hebrew, *bittul* means "self-nullification or selflessness," and to achieve this state, one is to engage in *"bittul hayesh,"* meaning the negation of one's selfhood (Wiederkehr-Pollack, 2007). Acts of selflessness and loving kindness outside the domain of administrative justice are morally and divinely mandated (Newman, 1990). Selflessness in this sense is an important feature of Judaism and is spoken of in terms of one's duties to God, described in the collective body of religious laws (*halakhah*) that are indicated in the central Jewish scriptures (the *Torah*, which later include the *Mishnah* and *Talmud*).

According to the central teachings of Judaism, all moral laws are either directly mandated or derived from the divine commandments of God. One is to follow God's commandments (*mitzvot*) to fulfill the eschatological purpose of creation, which is to attain individual salvation. However, in this endeavor, a person is not to be concerned solely about oneself, but rather to see himself or herself as insignificant (i.e., merely a tool) to fulfill their godly purpose in the world. In Hasidic Judaism, the idea of selflessness is developed through the realization of one's *bittul bim'tziut*, which in Hebrew means "an entity of nothingness" (Freeman, n.d.):

> The best time for *Hitbodedut* (i.e., contemplative prayers to G-d) is at night, when everyone is asleep. Ideally you should go to a place outside the city and follow a solitary path where people don't even go during the day. Empty your heart and mind of all your mundane preoccupations and then work to nullify all your negative traits, one after the other, until in the end you nullify all sense of self completely (*bittul*). First work on one character trait, then another and another, until you reach the point where you are free of any self-centeredness and any sense of independent existence (*bittul bim'tziut*, i.e., the paradoxical state of somethingness/nothingness).
> *(see Freeman, n.d.)*

Realization of this paradoxical state includes simultaneous awareness of both one's self and one's selflessness and is presented as an ultimate goal of Judaism (not only for the individual but also for the entire world). This is said to lead to *bittul bimtzius*, literally "utter and complete self-nullification," as one fulfills one's life's purpose:

> You must be as nothing in your own eyes. Then you will be worthy of attaining true self-nullification (*bittul*) and your soul will be merged with its root. The whole universe will be merged with you in your Source. You and everything with you will be merged in the Unity of God.
> *Rabbi Nachman of Breslov (1772–1810; from Likutey Moharan I, 52; quoted in Greenbaum, n.d.)*

Gershom Scholem's (1954) important treatise on Jewish mysticism quotes a notable passage attributed to the experience of a "Hasidic mystic" (but without indicating who said it or when):

> There are those who serve God with their human intellect, and others whose gaze is fixed on Nothing .... He who is granted this supreme experience loses the reality of his intellect, but when he returns from such contemplation to the intellect, he finds it full of divine and inflowing splendor.
> *Scholem (1954/1946, p. 5); also quoted in Stace (1960a, p. 107)*

This quotation fits nicely with what was said above about how the intellect must be realigned to cultivate the divine consciousness of the divine soul, which can only occur by achieving complete selflessness. Note that the "Nothing" that is to be fixated upon indicated in the quote above is capitalized, here referring to an experience where one is completely but only temporarily detached from the intellect. However, one then returns to their everyday cognizance, but now with a heightened awareness of the divine obtained from a profound experience of selfless spiritual transcendence.

While there are many writings and ritual practices pertaining to mysticism in the Jewish tradition, these aspects are often downplayed in mainstream Judaism. As David Baumgardt (1947) noted some time ago:

> Philosophers of Judaism have hesitated to admit that any kind of mysticism could be valued as a legitimate offspring of the purity of monotheistic Jewish thought. To all these interpreters of Judaism mysticism seemed disturbingly linked up with obscurantism, superstition, mythological fancies, with conjuring magic and questionable theurgical practices .... The Jewish interpreters of Judaism tried on the one hand to appease at all costs the world of modern rationalism—which, in its most marked tendencies, cannot be religiously appeased—and, on the other hand, they rejected the help offered by modern religious non-rationalism.
>
> *(Baumgardt, Commentary Magazine. July 1, 1947)*

This issue remains relevant today, especially in light of the enormous efforts made by Jews (and others) to assimilate to American culture and especially for European Jews who survived the Holocaust and came to the United States after World War II. As Stace (1960a) has noted, examples of Jewish mysticism found among Hasidic segments of Judaism are often regarded as somewhat heretical by Jewish orthodoxy. He states the following:

> Jewish tradition has always frowned on this kind of mysticism in which identity of even union, with God is claimed. Its emphasis is on the great gulf which separates God from his creation, so that a claim to a union or identity which negates the gulf generally seem objectionable to the religious Jew.
>
> *Stace (1960a, p. 106)*

Nevertheless, there is an ancient and long history of mystical Judaism, perhaps best exemplified by Merkaba, a school of Jewish mysticism (*c.* 100 BCE–1000 CE) that centered on visions such as those found in the book of *Ezekiel* (see Chapter 1 of *Ezekiel*) and on what is known as the "palace" (*hekhalot*) literature that contain stories about ascending to heavenly palaces and beholding the throne of God (see Scholem (1954) for a detailed discussion).

6 ISLAM, SELFLESSNESS AND UNITY

> I have become senseless, I have fallen into selflessness—in absolute selflessness how joyful I am with the Self.
>
> *Rumi (1207–73; from Mathnawi, Chittick, 1983, p. 174)*

In Islam, the earthly body and the divine spirit are seen as directly opposed. The self (or soul) is regarded as the mediator of this dichotomy, and this is where the worldly and the divine meet. In Arabic, *nafs* means the "self," "soul," or "being" (and is linguistically related to the term *nephesh* used in the Hebrew Bible). The self is said to have three levels, from the most base to the most divine (Sviri, 2002). The first level of self is driven by the desires, impulses, fears, and appetites, which compose the ego. At this lowest level, the self seeks only to protect its own self-interest. In the next intermediate stage, the self experiences inner struggle (*jihad*). Recognizing a higher reality, the self seeks to control its impulses and balance consciousness to fulfill the covenant with God. Here, one becomes aware of his or her own imperfections and seeks to improve oneself before Allah/God. In the final or third stage, through ritual, contemplation, spiritual practice, and religious teachings, the self finds tranquility in faith for Allah, and Allah is said to be revealed through various signs. Here, the self attains the highest possible spiritual level achieving peace with God (Sviri, 2002).

In embarking on this endeavor, selflessness (*ithar* in Arabic), in the sense of giving preference to others over oneself, is regarded as one of the required noble qualities. One who is selfless places concern for others above concern for one's own well-being. The selfless person is generous and freely lends aid and support to others. To be selfless, as some Sufis say, represents the "annihilation" of the self, which is expressed as *fana* in Arabic (Sviri, 2002). As the famous poet and Sufi mystic Rumi said,

> Purify yourself from the attributes of self, so that you may see your own pure essence.
> *Rumi (1207–73; from Mathnawi, Chittick, 1983, p. 173)*

To achieve *fana* means to break down the ego and reliance on the senses to find one's true essence. As Rumi stated,

> I will come to myself the instant I am obliterated and made selfless (*ithar*): I am complete only when outside of the five senses and four elements.
> *Rumi (1207–73; from Diwan i Shams i Tabriz, Chittick, 1983, p. 174)*

When this is achieved, one comes to realize the fundamental unity between God (Allah), creation, and the individual self (Harmless, 2008). This is considered an enlightened state of consciousness where one becomes cognizant of the intrinsic unity of God and existence. It is called *tawhid*, which means "oneness," and is the most important concept in Islam. It expresses that state of indivisible oneness and the monotheistic principle of Islam asserting the oneness of God/Allah. When *fana* is achieved, a state of pure consciousness (*baqaa*) is said to have been attained. It is the ultimate expression of the loss of self, a complete state of utter selflessness:

> In God there is no duality. In that Presence "I" and "we" and "you" do not exist. "I" and "you" and "we" and "He" become one .... Since in the unity there is no distinction, the Quest and the Way and the Seeker become one.
> *Mahmud Shabistari (CE 1320; quoted in Stace, 1960a, p. 106)*

III. SELFLESSNESS AS THE KEY TO TRANSCENDENCE

At this point, nothing is left but total oneness or singularity of awareness:

> When a man's 'I' is negated (and eliminated) from existence, then what remains?
> Consider, O denier.
> *Rumi (quoted in Nicholson, n.d.)*

In the mystical understanding of Sufi Islam, this is the complete surrender of oneself to God (Allah). In terms of neuroscience, this loss of self is likely articulated by a significant slowing or even shutdown of neural input into portions of the right parietal lobe and is worthy of more collaborative research between science and the humanities to better understand if such occurrences can be understood collectively.

7 SPIRITUALITY, AGNOSTICS AND ATHEISTS

The relationship between religion and spirituality appears to be shifting in new directions in modern times. Spirituality embraces emotional and cognitive experiences associated with a wide range of outlooks and beliefs. Today, we can speak of people who are "spiritual but not religious" and also refer to "nones," meaning those who do not claim affiliation with any particular religious group or faith tradition (see, e.g., Fuller, 2001; Mercadante, 2014). On the other hand, there are those who profess religious allegiance, for example, to avowedly monotheistic religions (e.g., Judaism and Christianity), but at the same time indicate that they do not believe in God and perhaps can be termed religious atheists. Yet even among confirmed nonbelieving atheists, the relationship between selflessness and spirituality is apparent. Sam Harris, a neuroscientist and modern spokesperson for atheism, discusses what he terms "the *intrinsic selflessness of consciousness*" in his book *Waking Up*, in a chapter entitled "The Riddle of the Self" (p. 82, original emphasis). In the next paragraph, he continues:

> My goal … is to convince you that the conventional sense of self is an illusion—
> and that spirituality largely consists in realizing this ….
> *Harris (2014, p. 82)*

While we are not in full agreement with Harris' arguments about selflessness and consciousness, we do note how frequently he connects selflessness together with experiences of spiritual transcendence and says that "the deepest goal of spirituality is freedom from the illusion of the self" (Harris, 2014, p. 123). Other modern notions that connect the loss of the sense of self with spiritual transcendence and come from outside of religious contexts are seen even earlier.

In the late 19th and early 20th centuries, the semantic field of spirituality started to extend beyond theological and ecclesiastical concerns.

By the mid-20th to late 20th century, modern notions of spirituality are undergoing a discursive shift and in terms of transcendence reveal a comprehensive experience of connectedness with other individuals, nature, or the cosmos often by virtue of a temporary loss of the regular sense of the immediate self. Conceptually, spirituality is developing into a new cultural category that at times defies the binary opposition between religious and secular, sometimes creating new cultural and social understandings that appear as neither religious nor secular. Thus, at times, the distinction between religious and atheist can become confusing, as in the French-Algerian novelist and playwright Albert Camus' assertion:

> It's true I don't believe in God, but that doesn't mean that I'm an atheist ....
> *Camus (quoted in Santoni, 2013, p. 142)*

While spiritual experiences may be deemed religious for some individuals, there are also nonreligious forms of spirituality recognized, and we can explore spiritual agnosticism or spiritual atheism. Nonetheless, the connection between selflessness and spirituality appears to persist, despite many conceptual complications. Einstein, who considered himself an agnostic but not an atheist, once said:

> The true value of a human being can be found in the degree to which he has attained liberation from the self.
> *Einstein (1949/1954, p. 8)*

Another famous agnostic, the astronomer Carl Sagan, said

> Science is not only compatible with spirituality; it is a profound source of spirituality. When we recognize our place in an immensity of light-years and in the passage of ages, when we grasp the intricacy, beauty, and subtlety of life, then that soaring feeling, that sense of elation and humility combined, is surely spiritual.
> *Sagan, 1996, p. 29*

It is notable that the experience of selflessness is promoted across a wide variety of religious traditions, despite their different historical and cultural brandings. Sometimes, this occurs by experiencing the complete loss of the sense of the self through merging with something larger than oneself, at other times by separating or emptying oneself from the totality of the universe. Sometimes, it is simply the experience of being in the presence of the divine however interpreted. Although the experience of selflessness is understood in myriad ways in different religious traditions, it is the experience of selflessness expressed in different forms that appears to provide a universal construct that is approachable within this manifest diversity from both spiritual and neuropsychological vantage points.

III. SELFLESSNESS AS THE KEY TO TRANSCENDENCE

Universal Neuropsychological Model of Spiritual Transcendence

> *The true mystic is not looking for peak experiences or altered states of consciousness. No, the genuine mystic is on a course of radical self-forgetting, self-surrendering, and self-transcending. (Griffin, 2003, p. 13)*

© 2019 Elsevier Inc. All rights reserved.

1 INTRODUCTION

The previous chapters have laid the foundation for a neuropsychological model of spiritual transcendence that applies universally to individuals from all cultures and religions. In developing our model, we build upon the efforts of others who have previously proposed related models of spiritual transcendence and have also focused on the relationship between aspects of the self and spiritual transcendence (Austin, 2009; McNamara, 2009; Newberg, d'Aquili, & Rause, 2001; Roberts, 2005). We reiterate that our neuropsychological model of spiritual transcendence is pared down in terms of neurological networks and spiritual concepts when compared with these previous models, as we focus on the association between a specific neuropsychological process (i.e., self-orientation) and a specific spiritual experience (i.e., spiritual transcendence). Our emphasis is on spiritual experiences that involve a diminished sense of self and an associated emotional experience of unity with the cosmos or the divine, however conceptualized (e.g., religious, spiritual, transcendent, and mystical).

First, we focus on the neuropsychological processing of spiritual transcendence, including how the brain integrates sensory and mental experiences into a "sense of self." We suggest that this sense of self can be inhibited leading to both a decreased sense of the self and simultaneously an increased experience of intense emotional connection with the cosmos/divinity (i.e., with something greater than oneself). We then describe the bonding and affective aspects of spiritual transcendence by focusing on the relationship between certain neuropeptides and neurotransmitters (i.e., oxytocin and dopamine) and their effects on spiritual disposition and religious experience. We conclude by discussing how sociocultural and religious factors can influence both the experience and interpretation of occurrences of spiritual transcendence.

This last segment is important as we appreciate that "self" is not just a solitary or independent construct but rather refers to an experience of awareness that is in constant interaction with the (internal and external) world and that "self" is dynamic and not static. This means that "self" includes both the individual (e.g., I and me) and also one's connections with others (e.g., family, religion, culture, nation, and cosmos). Humans are inherently social beings who engage collectively in ongoing social interactions. These interactions continually influence the construction of the self from neuropsychological, neurophysiological, and social perspectives. This is the hardest part of our model to elucidate, but we realize that it must be addressed in an initial manner.

2 THE NEUROPSYCHOLOGY OF SPIRITUAL TRANSCENDENCE

In simple terms, our model suggests that spiritual transcendence is, in part, a process of becoming selfless, often described as the "loss of the sense of self." We argue that neuropsychologically, the right hemisphere/parietal lobe integrates different sensory experiences and thoughts into a unified sense of self, and when this process of integration is diminished, it is possible to have spiritually transcendent experiences. This is not a new idea. It was described from a neurophysiological perspective by Eugene d'Aquili and Andrew Newberg in their book *The Mystical Mind* (1999) and later reiterated (with help of Vince Rause) in *Why God Won't Go Away* (2001). However, we are establishing this relationship more broadly by building a wider model that connects the neuropsychological loss of self with increased spirituality not only in terms of religious practices and experiences but also in cases of brain injury or through methods of neural inhibition that specifically cause reduced right hemisphere and parietal lobe functioning.

In our previous articles on the neuropsychological aspects of spiritual transcendence, we presented a relatively simple model that proposed that the right hemisphere/parietal lobe was associated with "self-orientation" (also described in the literature as self-awareness, self-consciousness, self-referencing, and self-focus), arguing that a significant reduction in self-orientation allows one to connect with things beyond the self (e.g., divinity, nature, and cosmos). However, while researching information for this book, we have become convinced that what is needed to enhance and broaden this understanding is a greater clarification of what constitutes "self-orientation" and how its reduction can lead to experiences of spiritual transcendence.

3 PROCESSING A SENSE OF SELF

In order to better understand how a diminished sense of self is related to spiritual transcendence, it is first necessary to understand how the sense of self operates as a neuropsychological process. A central component of our model suggests that the self is a neuropsychological process that is constantly being adjusted and readjusted, rather than something that is a permanent or fixed neuropsychic entity. When conceptualized as a process, it is possible to understand how the "self" can be diminished progressively to the point where experiences of spiritual transcendence occur.

As the studies described in Chapter 4 illustrate, the brain perceives different sensory (i.e., sight, sound, touch, taste, smell, and spatial positioning) and mental experiences (i.e., thoughts), but it is the right hemisphere/parietal lobe that *integrates* these experiences into a unified "sense of self." That is, the right hemisphere integrates these experiences so that they are perceived from a common point of perception (i.e., "my" experiences).

If the sense of self is a neuropsychological process, it is appropriate to assume that it may be altered (i.e., decreased and distorted). For example, if the brain receives conflicting or ambiguous sensory or mental information, individuals may perceive a distorted sense of self such as perceiving that a rubber hand is part of their self, that an amputated limb has been reattached to the body, or that the mental self is located in a spatial position that differs from the bodily self (e.g., out-of-body experiences). In these situations, it is essential to note that the ability to perceive different sensory experiences is intact, but the process of *integrating* these sensory experiences into a unified sense of self is distorted. These examples help show how the ability to create a sense of self works as a neuropsychological process that can be altered.

However, our model does not suggest that it is simply a *distorted* sense of self that is associated with transcendent experiences, but rather that a significantly *decreased* sense of self is necessary for spiritual transcendence. Individuals with a distorted sense of self still have a sense of self (i.e., "my" rubber hand, "my" amputated arm, and "my" body below me), although it is an inaccurately perceived sense of self. In contrast, individuals with a greatly decreased sense of self are able to accurately perceive sensory experiences but just without the more usual perspective of the self or the sense of self that they had previously become accustomed to experiencing. For example, individuals with mirror misidentification disorder have intact sensory experience (i.e., they see a person reflected in the mirror), but they cannot recognize the reflection as their own because there is no longer a neuropsychological process by which to integrate this information into a sense of self. It could be said that one cannot recognize the "me" in the mirror if there is no "I" from which to perceive the image.

In order to understand how the sense of self functions as a neuropsychological process, it is helpful to compare it with the manner by which other neuropsychological processes operate. For example, research is increasingly showing that the brain processes a "sense of memory" (and not specific, unchangeable memories) that can change over time (for an excellent description of the process of memorization, go to www.npr.org/templates/story/story.php?storyId=1134345). As such, the process of memorization can be *decreased* leading to amnesia (i.e., the loss of memories) or *distorted* leading to confabulation (i.e., inaccurate memories and recollections as in Korsakoff syndrome). The processing of creating a sense of self works similarly in that it can be *decreased* leading to conditions like

III. SELFLESSNESS AS THE KEY TO TRANSCENDENCE

asomatognosia (i.e., the loss of the physical sense of self) or *distorted* leading to out-of-body experiences (i.e., an inaccurate sense of the location of the physical self).

We argue that similar decreases in the manner by which the sense of self is processed can serve as a neuropsychological foundation of spiritual transcendence, but note that such experiences are not pathological or indicative of neurological impairment. Instead, they are experiences that can occur when individuals have learned to minimize the sense of self through religious rituals or other behavioral practices or when they occur spontaneously by other means (e.g., through divine intervention and harmonization with nature).

4 SELFLESSNESS AND UNITARY CONSCIOUSNESS

Certainly, not all experiences of selflessness lead to spiritual transcendence. If this was the case, individuals with neurological disorders of the self would experience different degrees of near-constant transcendence. This raises that question as to what other neuropsychological processes besides selflessness are necessary for the experience of spiritual transcendence. Based on the phenomenological descriptions of transcendent experiences throughout history, it does appear that they are associated with both increased selflessness and an associated sensation of being an undifferentiated unity (also reported as unitive experiences, unitary consciousness, the universal form, etc.).

Just as we have worked to provide a more refined explanation for how the brain creates a sense of self, we also realized the need to develop a better neuropsychological explanation for the sense of universal connectedness that is often reported from a wide variety of experiences of spiritual transcendence. As has been noted throughout this book, it is difficult if not impossible for individuals to accurately or fully describe the feelings and perceptions connected to their experiences of spiritual transcendence if they no longer have a sense of self from which to report the experiences. Nonetheless, this has not stopped people from doing so throughout history, and the descriptions made by many different individuals and as recounted in Chapters 2 and 6 provide useful insights into the manner by which a reduced sense of self may lead to an experience of universal connectedness.

People describe their individuality dissolving and fading away, feeling themselves becoming undivided and in unity, as their sense of self becomes absent and they "became one," as Jane Goodall says of her experience, "with the spirit power of life itself" (Goodall & Berman, 1999, p. 173). It is a pattern of description that occurs repeatedly in different times, places, and contexts, and we believe this is not coincidental. We suggest

III. SELFLESSNESS AS THE KEY TO TRANSCENDENCE

that by decreasing the sense of self, one becomes open to experiencing perceptual and emotional interconnections beyond the self. We suggest that when one does not perceive experiences as relating to one's self, it is possible to perceive experiences as relating to all interconnected beings. What is necessary is to develop better explanations for the increased sense of connection/bonding beyond the self and the intense positive emotions associated with spiritual transcendence that are often life-changing.

5 OXYTOCIN, EMOTIONAL BONDING, AND SPIRITUAL TRANSCENDENCE

Our prior articles on a universal neuropsychological model of spiritual transcendence have focused almost exclusively on cognitive and perceptual processes associated with spiritual transcendence (i.e., decreased perception of the self). However, whereas inhibited right hemisphere/ parietal lobe functioning may lead to a decreased sense of self, this cannot entirely account for the sense of intense emotional bonding and connection associated with spiritual transcendence. Specifically, a diminished sense of self is different from an intense emotional connection beyond the self (e.g., unity, awe, and wonder), as they are not only distinct but also related. A more comprehensive explanation for the experience of connectedness associated with spiritual transcendence is needed. In this regard, recent research in neuroendocrinology is beginning to show that the activity of specific hormones may partially explain the unitive experiences that are commonly described with self-transcendence.

For example, oxytocin, a hormone synthesized in the hypothalamus of the brain, can be measured through the analysis of blood plasma or by testing saliva. Initially, oxytocin was recognized primarily as the hormone that facilitated the emotional bonding that occurs between mothers and their newborns. It has been well established that stimulation of the nipples in women causes the release of oxytocin, effects that are important in mother-infant bonding during breastfeeding. More recent research is also showing that oxytocin influences how the brain processes the sense of self, reducing focus on the self, and that it may be connected with experiences of self-transcendence.

For example, a study by Liu et al. (2013) found that the release of oxytocin significantly weakened self-referential processing, with the right parietal lobe showing maximum deactivation. These findings have direct relevance to our model of selflessness occurring related to decreased functioning of the right parietal lobe, which leads to a decreased sense of the self. Additional research suggests not only that oxytocin can promote a reduced focus on the self but also that it may stimulate a heightened spiritual disposition, including a sense of connection with the world or with a higher power (Van Cappellen, Way, Isgett, & Fredrickson, 2016). Importantly, these feelings of increased spirituality were seen to persist even 1 week after the initial administration of oxytocin, suggesting that the spiritual orientation engendered by oxytocin was deeply embedded and left lasting impressions. Interestingly, even when controlling for possible cultural influences, such as individuals' involvement in organized religion, studies have shown that higher levels of oxytocin are significantly associated with producing higher levels of spirituality (Holbrook, Hahn-Holbrook, & Holt-Lunstad, 2015; Kelsch et al., 2013).

Taken together, these studies suggest that the stimulation of oxytocin may be associated with perceptual processes that reduce focus on the self, thereby increasing the potential for experiences of emotional bonding (i.e., the unitive experience) associated with self-transcendence. Oxytocin is implicated not only in maternal-infant bonding but also in sexual/mate bonding that may occur during orgasm in both men and women (Georgiadis, Reinders, Paans, Renken, & Kortekaas, 2009), when one may merge emotionally with another. In general, this area of research reinforces our view of the interconnections that can develop between the loss of the sense of the self and experiences of spiritual transcendence. When potent experiences of selflessness occur in the framework of religious and spiritual contexts, strong emotional attachments with profound understandings of the truth and connections with the divine become possible, transcendent experiences that remain influential for these individuals for the rest of their lives.

6 DOPAMINE AND SPIRITUAL EXPERIENCE

Whereas oxytocin may help explain the experience of intense "connection" (i.e., bonding) with the divine that is reported during spiritually transcendent experiences, specific neurotransmitters are being shown to be associated with the positive emotional experience of transcendence. In comparison with other animals, the human brain is heavily invested with reward centers. The neurotransmitter dopamine plays an important role in producing our feelings of reward through its effects on various parts of the brain, especially in areas of the prefrontal cortex. Brain regions associated with rewards play an important role in the release of the "feel-good" neurotransmitter dopamine. Dopaminergic neurons have long been associated with the reward and pleasure systems of the brain, and almost all known addictions work by prolonging the influence of dopamine producing heightened experiences of reward (McNamara, 2009). However, it has also been noted that spiritual feelings that create happy sensations and rewarding feelings may also have a significant dopaminergic component.

In a one recent neurophysiological study involving devout Mormons, researchers found that certain brain regions were activated consistently when these individuals reported having strong positive spiritual feelings (Ferguson et al., 2018). It appeared that their dopamine reward regions were activated to produce these effects. More generally, the neuropsychologist Patrick McNamara has indicated that boosting the activity of dopamine on the right side of the brain intensifies the spirituality (what he calls the "god effect") of people already attracted to religious ideas (McNamara, 2014). However, while these occurrences reflect heightened religious experiences, they do not appear to be instances of spiritual transcendence.

These findings are also consistent with the research of the geneticist Dean Hamer who suggests that the propensity for spirituality is genetically inheritable and specifically related to a gene associated with the production of dopamine (Hamer, 2004). Thus, the intense, positive emotions described during spiritual transcendence may be related to the release of dopamine that is typical for any rewarding experience. Alternatively, there may be a genetic predisposition for spiritually transcendent experiences that are also naturally rewarding.

7 RELIGION AND THE LOSS OF SELF

Theoretically, if spiritual transcendence is neurologically based as we contend, it is appropriate to assume that all individuals will report the phenomenological aspects of these experiences in generally similar ways (e.g., selflessness with universal consciousness). However, these experiences have been interpreted quite differently throughout history. Although our model focuses on the neuropsychological aspects of spiritual transcendence, there is a need to explain how these universally similar neuropsychological experiences are given such different interpretations by individuals from different faith traditions.

While we propose that the neuropsychological processes engaged during potent experiences of spiritual transcendence are similar for all individuals (i.e., a decreased sense of self, increased sense of bonding/connection, and intense positive emotional experience), the manner by which these experiences are interpreted and understood is based on cultural and religious contexts and backgrounds. In fact, religious explanations of what has occurred during these experiences often vary greatly. As the medieval Christian theologian Meister Eckhart is reputed to have said, "theologians may quarrel, but the mystics of the world speak the same language" (Easwaran, 1996/1989, p. 24).

Our ideas are also consistent with those of William James (1902), Harvard professor and founder of the study of the psychology of religion, and Walter Stace (1960a, 1960b), a British philosopher who wrote extensively about mystical experiences. Both of these scholars suggested that powerful spiritual experiences have a similar biological basis but are interpreted differently based on sociocultural and religious backgrounds. In his edited book, *The Teachings of the Mystics* (Stace, 1960b), Stace explained how similar sensory experiences could be interpreted differently:

> On a dark night out of doors one may see something glimmering white. One person may think it a ghost. A second person may take it for a sheet hung out on a clothesline. A third may suppose that it is a white-painted rock. Here we have a single experience with three different interpretations. The experience is genuine, but the interpretations may be either true or false. If we are to understand anything at all about mysticism, it is essential that we should make a similar distinction between a mystical experience and the interpretations that may be put on it either by mystics themselves or by nonmystics. For instance, the same mystical experience may be interpreted by a Christian in terms of Christian beliefs and by a Buddhist in terms of Buddhistic beliefs. *(Stace 1960b, also quoted in Pike, 1992, p. 98)*

We expand upon this model by suggesting that the "genuine" experience (as per Stace) is based in neuropsychological processes that involve both a reduced sense of self and an increased sense of connection and that are then understood through respective sociocultural contexts and religious orientations.

8 LANGUAGE AND SPIRITUALITY

One way to understand the manner by which neuropsychological processes and cultural influences interact to create experiences of spiritual transcendence is to make an analogy with the way that people develop language capabilities. All individuals are genetically predisposed to develop language skills that are neurologically hardwired primarily in the left hemisphere (for about 90% of people). However, the specific language one learns to speak while a child is also determined by cultural and environmental factors.

The manner by which spiritual transcendence develops operates in a similar manner. We are all programmed genetically to develop the same neurological structures and networks necessary for spiritual transcendence (i.e., right hemisphere-based selflessness). However, this spirituality is influenced by the sociocultural contexts in which we are raised. Thus, an individual is likely to interpret spiritually transcendent experiences involving the neurophysiological loss of the sense of self according to the belief (or nonbelief) system in which they are raised. We argue that this is how such similar neuropsychological experiences (i.e., selflessness and unitary consciousness) can be explained so differently in experiences like unio mystica in Christianity, enlightenment (*bodhi*) in Buddhism, full contemplative awareness (*samadhi*) in Hinduism, complete nullification of self (*bittul hayesh*) in Chasidic Judaism, or selfless pure consciousness (*baqaa*) described in Islam.

9 CONNECTING NEUROSCIENCE WITH RELIGIOUS UNDERSTANDINGS

While writing this book, it became apparent that there are different and conflicting theories about the nature of the self that we need to address in our model. For instance, several scholars have noted that spiritual transcendence can be expressed as feeling that the self is devoid of attributes (Hood, 2002) or that it may involve a sense of self that has become unified with a greater whole (Sullivan, 1995). Some have tried to argue that these different perspectives on the loss of the sense of the self must be opposites (Baumeister & Exline, 2002; see discussion in Simpson, 2014). However, we believe these are interrelated understandings of how selflessness can be achieved. Whether couched in "self-as-all" or in "no-self" language (Simpson, 2014, p. 461), we argue that the neuropsychological processes that are occurring are similar. Our contention is that when the processes of integrating sensations and experiences into a sense of self are diminished significantly (i.e., no self), it becomes possible to experience a greater range of sensations and experiences that go beyond the sense of the self (i.e., self as all). We contend that these are not two separate experiences, but rather one experience (no self) that may lead to the other experience (self as all).

This seemingly polar orientation of the loss of the sense of self is already expressed vividly in divergent religious understandings of the nature of the self and the attainment of salvation. This contrast is seen, for example, when viewing religious orientations to selflessness in terms of the Buddhist realization of enlightenment (*bodhi*) and "no self" (i.e., *anatman*), in comparison with the Hindu understanding of the self as something that is eternal (i.e., *atman*) and if comprehended properly is realized as being equivalent with universal totality (i.e., *Brahman*). Thus, we see theological conceptions of the nature of the self that are fundamentally different, even in religious traditions that are historically and theologically closely related, and where both ascribe to the idea that reincarnation (*samsara*) must be overcome to attain ultimate salvation. Here, the potential for salvation can be expressed through religious understandings that describe the highest attainments of spiritual transcendence as involving either the total eradication of the (ultimately false) self in Buddhism or as the unity of the (eternally existing) self with the cosmic totality in Hinduism. It may be that these experiences are not polar opposites, but in fact are related aspects of the same neurophysiological process.

10 CULTURE AND NEURAL PROCESSING OF THE SELF

In addition to creating deep insights about how we experience religion and spirituality, we also note that culture creates interpretive frameworks that can have consequences in terms of neurophysiological development, functioning, and neuroplasticity. That is, research is suggesting that culture not only affects the manner by which neuropsychological experiences are interpreted but also affects the manner by which neuropsychological processes of self-orientation develop within the brain. As such, individuals may be predisposed to interpret experiences of selflessness according to the culture in which they were raised and the manner by which the self is conceptualized within one's culture.

For example, several studies show how different cultural conceptualizations of the self actually influence the manner in which the self is neuropsychologically processed. In general, it appears that Western cultures tend to emphasize the importance of the individual who is separate from others (i.e., individualistic orientation), while Asian cultures often emphasize the importance of the individual within the context of the family, social group, and community (i.e., collectivist orientation). Research suggests that the brain processes information differently based on whether one is culturally oriented toward individuality or oriented toward the wider context of their community interconnections.

For example, in a study of college students, English-speaking Westerners (e.g., British, American, Australian, and Canadian) exhibited consistent neural differences in processing images of themselves and of others, when compared with Chinese-speaking East Asians (Zhu, Zhang, Fan, & Han, 2007). Additional research has found that a Western *independent* "self" orientation contrasted with an East Asian *interdependent* "self" alignment and could reflect the operation of different cultural values pertaining to individualism and collectivism, respectively (Chiao et al., 2009). Neuroimaging studies have also shown the existence of distinct neural substrates for self-related processing for nonreligious Chinese compared with Chinese Christians (Han, Gu, et al., 2010; Han, Mao, et al., 2008). In other research, Han Chinese participants showed stronger neural "self-processing" in contrast to "other processing" when compared with Tibetan participants who exhibited a different neural self-reference pattern. These differences may reflect neurological functions operating more in accord with the philosophical and religious views of the false sense of self that are articulated in Tibetan Buddhism and Tibetan culture (Wu, Wang, He, Mao, & Zhang, 2010).

11 SPIRITUAL TRANSCENDENCE, THE UNAFFILIATED, AND NON-BELIEVERS

Neuropsychological models of spiritual transcendence are needed for individuals who fall along a wide range of beliefs and non-beliefs, practice and nonpractice of religion, and the many different permutations that are found between these categories. Throughout this book, we have suggested that individuals who do not ascribe to any specific religious creed or affiliation can also have spiritually transcendent experiences. These identities are increasingly important to consider, as over one-quarter of individuals in the United States now report being spiritual but not religious (Pew Research Center, 2017), and the number of people indicating that they identify as atheist is also growing, especially among younger segments of the population (Pew Research Center, 2015).

When researching information for this book, it became clear that there are few if any adequate neuropsychological models of spiritual transcendence for people who might ascribe to what could perhaps be called different forms of "spiritual atheism." Sam Harris (2014) is a good example of a self-identified skeptic and atheist, a long-term practitioner of meditation (which he terms a spiritual practice), which is to be used as a way "to stand free of the juggernaut of the self, if only for moments at a time" (p. 11). Harris advocates what he calls "a rational approach to spirituality" (p. 10) to explore the "phenomenon of self-transcendence" (p. 18), which he sees as "the deepest goal of spirituality" and the way to attain "freedom from the illusion of the self" (p. 123). We believe Harris presents a good example of an atheist spirituality that fits well with our model of spiritual transcendence and the neuropsychology of the loss of the sense of self. The point here is that neuropsychological models of spiritual transcendence are needed for exploring a wide range of beliefs and nonbeliefs and the increasingly changing conceptions of the interrelationship between religious and spiritual identity, where a wide variety of changing permutations are occurring (see e.g., Pew Research Center, 2015).

It is also possible that "flow" states (discussed briefly in Chapter 1) and spiritual transcendence are distinct but related experiences that are both united by a decreased focus on the self and the experience of connection or unity with objects or activities beyond the self. Research suggests that individuals can have selfless, transcendent experiences in religious or nonreligious contexts, such as when hiking in the forest (Williams & Harvey, 2001) or viewing the earth from space (White, 1987; Yaden et al., 2016). Astronauts often describe gazing back upon the earth as "truly transformative experiences involving senses of wonder and awe, unity with nature, transcendence, and universal brotherhood" (Vakoch, 2012, p. 29). One astronaut, Edgar Mitchell, reported an "overwhelming sense of oneness and connection ... accompanied by an ecstasy ... an epiphany" (Hunt, 2015, p. 77). Whereas some astronauts described this feeling of connection and unity in religious terms, others did not. The point is that the study of many different kinds of transcendent states may be an excellent model from which to understand the "spiritual" transcendence of individuals who do not ascribe to any religious identity.

12 SCIENCE AND THE HUMANITIES WORKING TOGETHER

Although the model of spiritual transcendence that we have presented is relatively simple in terms of neurophysiological networks and theological terminology, it provides a basic framework from which to better understand and explain the nature of spiritual transcendence. This model presents a great opportunity for the neurosciences and the humanities to work together in order to better explore and understand the dynamics of the self, selflessness, and spiritual transcendence. Developing this area of knowledge is particularly important as it has potential practical applications for enhancing religious and nonreligious experiences and understandings and improving health and human social relations.

III. SELFLESSNESS AS THE KEY TO TRANSCENDENCE

APPLICATIONS OF SELFLESSNESS

Building Bridges Between Neuroscience and the Humanities

OUTLINE

Science without religion is lame, religion without science is blind. **(Einstein, 1930)**

© 2019 Elsevier Inc. All rights reserved.

1 SCIENCE AND THE HUMANITIES

Religious traditions have hailed the virtue of spiritual experiences for thousands of years, and the neurosciences are beginning to develop an improved comprehension of such experiences. To advance the understanding of spiritual experiences, a working collaboration between the humanities and the sciences is needed. By working together more productively, the humanities and sciences can increase the understanding of the nature of selflessness and spiritual transcendence and develop important insights and tools that can help humanity adapt to a rapidly changing world. Instead of an ongoing battle between "believers" and "reductionists" as has often been engaged (see Cho & Squier, 2008; Slingerland, 2008), with a collective endeavor, we believe it is possible to develop different ways in which selflessness can be applied to enhance spiritual experiences and to promote better health and functioning in our lives.

We suggest that it is essential for various academic disciplines and professions to work together in order to better understand and enhance the potentialities of spiritual transcendence. This will require collaboration between the sciences and humanities that includes a wide range of academicians and should also include religious leaders and practitioners. As we have argued throughout the book, we believe that a science/humanities collaboration in the study of spiritual transcendence may be achieved by focusing on "selflessness" from the different perspectives of various disciplines. Without this integration, neuroscientists will likely continue to focus on neuroanatomical regions and neurophysiological mechanisms, and theologians and philosophers will likely provide sophisticated theories about the nuanced aspects of transcendence according to different faith traditions. However, by focusing on selflessness and working together, it may be possible to enhance experiences of spiritual transcendence for persons from numerous faith traditions and also for nonbelievers. Such an endeavor would also encourage conversations between faith traditions and nonbelievers, rather than just focusing on differences, and may encourage not only improved tolerance but also useful dialogue and even a sense of pluralism between faith and nonfaith orientations.

The possibility of this occurrence remains complicated in many ways. For example, in theology, religious studies, and philosophy, there is an ongoing theoretical argument about the nature of religious experience and by implication spiritually transcendent experiences. This discussion includes differences of opinion about the nature of religious experiences, for example, whether religious experiences are even comparable due to their contextual embeddedness and whether religion is sui generis (a thing in itself) or always determined by attribution (Taves, 2009). We believe a focus on selflessness may help progress such issues and move conversation forward productively.

Our interest in spirituality and selflessness has been guided by the ideas of earlier scholars who have explored religion, self and selflessness, and spiritual transcendence, such as William James (1902) and Walter Stace (1960a, 1960b), and also by later researchers who incorporated the tools of modern neuroscience including Andrew Newberg, Eugene d'Aquili, and Vince Rause (2001), and Patrick McNamara (2009). Based on our different academic outlooks and ongoing interdisciplinary collaborations, we provide general suggestions for future research not only on the neuropsychology of spiritual transcendence but also about phenomenological orientations. We further suggest the need to identify the specific nature by which a sense of self is processed within the brain as part of this agenda and also how these processes can be inhibited leading to selfless experiences of spiritual transcendence. In this agenda, we also see the importance of investigating the manner by which neuropsychological experiences of selflessness are interpreted differently, based on religious and nonreligious beliefs, as well as cultural orientations.

Science is needed to answer questions pertaining to the neuroanatomical, neuropsychological, and neurophysiological mechanisms of spiritual transcendence, and the social sciences are needed to explore the psychological, cultural, and social mechanisms by which spiritually transcendent experiences lead to improved health, human flourishing, and social relationships. However, the humanities are also essential, not only to address the nature and meaning of human relationships to the divine but also to help other areas of investigation deal more directly with the human experience of spirituality, particularly through first-person accounts. Without better integration between the sciences and humanities, future research in this area will remain, in the words of Albert Einstein, "lame" and "blind," and progress will remain limited.

2 NEUROTHEOLOGY AND ITS LIMITATIONS

It has been suggested that neuroscience and theology can collaborate directly through a combined field of neurotheology. Andrew Newberg has been the strongest advocate for a "neurotheology" that is based in part on research using different forms of neuroimaging to map differential levels of brain activity in individuals performing specific religious rituals (e.g., meditation or prayer). This type of research has produced important and interesting results, but how should it be interpreted and what are its wider implications? On the one hand, this type of research shows that science cannot simply dismiss religious and spiritual experiences as mere delusions that are invented and propagated, if there are consistent cerebral bases seen for these experiences. On the other hand, neurotheology becomes a dangerous form of scientific reductionism if religious and spiritual experiences (despite their varied first-person interpretations) are reduced solely to the operation of neural processes. Newberg argues that "neurotheology" brings science and religion together (Newberg, 2010, 2018; Newberg, d'Aquili, & Rause, 2001), but this contention remains problematic. Acknowledging that there are biological bases for human spirituality does not mean that these experiences can be fully explained in terms of neural activity. Larger aspects of context, sociality, and culture, as well as more direct issues concerning embodiment (as the brain is part of the body), are not merely secondary aspects in a brain-based understanding of religion. As the religious studies scholar Manuel Vásquez (2011) has noted:

> Religion is not simply a set of neural and biochemical processes that take place inside the skull. At best, we can claim that certain genetic and neuronal configurations may predispose some individuals toward experiences and moods that are lived as religious. This predisposition is mediated by social discourses and practices. (*p. 192*)

The broader question is how to involve academic scholars, religious leaders, religious practitioners, mystics, and nonbelievers in the study of spiritual transcendence. There is no simple answer, and as Newberg and his colleagues (Newberg, Alavi, et al., 2001; Newberg, d'Aquili, et al., 2001) pointed out long ago, the lack of response by religious leaders to the fledging neuroscientific study of religious experience through neuroimaging studies was in itself surprising. This nonconversation appears, for the most part, to be ongoing.

An ongoing problem with neurotheology, as another scholar of religion has pointed out, is that while Newberg "repeatedly calls for an equal and mutually respectful partnership between religious and scientific perspectives ... it is clear that the real engine of neurotheology is brain science" (Barrett, 2011, p. 134). We too not only have struggled with this limitation but also believe that we have taken a step in the right direction by focusing on the neuropsychological and religious experience of selflessness as a linking component, including the incorporation of religious and nonreligious perspectives about spiritual transcendence. Thus, our model does not advocate neurotheology directly, as it acknowledges that context and interpretation are driving factors in what constitutes a spiritually transcendent experience. We understand that the problem of reductionism often occurs when the sciences and humanities meet, and we are suggesting, for the time being, a more noneliminative reductionism, which clearly recognizes that there are complex emergent human structures of meaning, like religion (see Slingerland, 2008, for an extended discussion). We encourage scholars from the humanities to work together with neuroscientists and cognitive scientists, to create insights and understandings that recognize not only neurophenomenology and neuroplasticity but also "communities of discourse and practice" that also operate in producing understandings of spiritual experiences (Taves, 1999). Some progress in this direction has occurred, perhaps most predominately displayed in the interface of research into neuroscience and Buddhism.

3 NEUROPHENOMENOLOGY AND SPIRITUALITY

Advances in neurophenomenology and the study of spiritual experience have been most pronounced in Buddhist studies. One notable instance of a more integrative approach to spiritual transcendence is seen in the work of James Austin on the neurology of spiritual experience and its relationship to the processing of the sense of self viewed from a scientific and Zen Buddhist perspective. Austin is a neurologist, an accomplished researcher, and also an advanced practitioner of Zen Buddhism. He has written several important books on the neuroscience of meditation in relation to transcendent states of Zen Buddhism (e.g., Austin, 1998, 2006, 2009), working from the position of not only a neuroscientist but also as one who has an insider's grasp as a devout and highly experienced practitioner. In his various analyses, Austin provides ideas on how the brain neurologically processes a sense of self and how losing this sense of self may be associated with transcendent experiences. His writings are exceptional and neurologically complex and may be most beneficial to our model of spiritual transcendence by describing different neural networks associated with different senses of the self and nonself.

In particular, he has described the manner by which the brain processes spatial information from two opposing perspectives, from the subjective position of oneself and separately from an objective orientation of an object outside the self. The subjective self sees an external object from a self-centered orientation (i.e., the perspective of looking outward from the body), which is referred to as "egocentric processing." In contrast, in "allocentric processing," an object is perceived as if the object being observed is itself doing the seeing and is looking back toward the one who is seeing. While the sense of self is predominately built on the former (i.e., egocentric processing), Austin's contention is that through the long-term practice of Zen meditation, one becomes increasingly adept at deactivating both (i.e., egocentric and allocentric) neural pathways. By decreasing the neurological processes of the subjective sense of self and the objective sense of self, it is possible to progressively lose one's false sense of the self entirely, which is the core tenet of Buddhist enlightenment. Austin's later books build on these neurological insights and Buddhist religious orientations, explaining how to engage in Zen meditation practice to develop a fuller understanding of "lived in" Zen and showing the potential of integrating insights from neuroscience with religious perspectives.

This type of integrative approach appears to be the exception rather than the norm in studying the neuroscience of meditation and selflessness. Other related developments include the work of researchers who work directly with experienced Buddhist meditators in neuroscience laboratories and have trained these highly proficient practitioners in phenomenological methods of introspection, in order to create a more sophisticated and integrated platform for developing more productive dialogue between practitioners and scientists (see, e.g., Harrington & Zajnoc, 2003). Neuroscience researchers have also been working with meditators' first-person accounts of their ritual practice and spiritual experiences to create a more detailed inventory of different Buddhist meditation practices (Lutz, Dunne, & Davidson, 2007). These procedures and methods will help deal with the problem of seeing all forms of meditation as directly comparable, as if one form of "meditation" practice is automatically comparable with any other. This type of work represents some of the initial steps in the direction of better integrating science and the humanities.

Notably, many of the neuroscientific studies of meditation conducted by neuroscientists have focused on Buddhist types of meditative practices and especially on "mindfulness" meditation. This body of research is growing and has represented an important advance in understanding the relationship between meditation and brain function in general. What is curious is the dissociation seen in many of the studies of Buddhist meditation in regard to the self and selflessness, which is central to Buddhist religious perspectives. As Goleman and Davidson (2017) have recently noted,

> Lessening the grip of the self, always a major goal of meditation practitioners, has oddly been ignored by meditation researchers, who perhaps understandably focus instead on more popular benefits like relaxation and better health. And so, a key goal of meditation—selflessness—has only thin data, while other benefits, like health improvements, are heavily researched … *(p. 161)*

4 NEUROSCIENTIFIC AND NEUROPSYCHOLOGICAL MODELS OF THE SELF

Recent theories are suggesting the existence of different neurophysiological mechanisms for the manner by which individuals perceive different aspects of the self. Although they differ, all suggest the need to study different forms of consciousness as they relate to how a sense of self is generated. In particular, neuroscientific research is beginning to assist in differentiating between the experience of consciousness that occurs with self-awareness (i.e., self-consciousness) and experiences of consciousness without self-awareness (i.e., unitary or selfless consciousness).

In his book *Self Comes to Mind* (2010), the neurologist Antonio Damasio has provided some possible insight into these differences through his work with patients with unusual neurological conditions, such as anencephaly. These individuals have unusually small brains with minimal cerebral cortex (i.e., they primarily have only brain stem and limbic system structures). Based on his observations, Damasio has proposed that there are two basic processes associated with a sense of self, including a "protoself" and a "core self." He suggests that the protoself exists in both humans and animals and is associated with a primitive sense of self that perceives internal conditions and states. In contrast, the core self corresponds to a neurophysiologically transient process generated continuously relative to any object with which an organism interacts. He believes that when an organism attains awareness of the feeling that its protoself is being affected by its sensory experiences, then "core consciousness" (i.e., the "core self") arises. But core consciousness reflects only the present moment, the here and now, and it does not require language or even memory. However, these "core consciousness" experiences and other neurophysiological operations produce what Damasio calls "extended consciousness." Here, one's awareness of self is interconnected with the sense of other beings in a reflective and resonating manner. This suggests that extended consciousness relates to the integration of basic sensory experiences into a sense of self, which we believe may become reduced significantly during experiences of spiritual transcendence.

5 THE EXPERIENCE OF THE "SELF"

In order to better understand the nature of spiritual transcendence, we suggest that it is necessary for disciplines to contribute collectively to develop a better understanding of the nature of the self in respect to selflessness. In simple terms, there is a need to understand the experience of selflessness (i.e., no self) and how it relates to spiritual transcendence. This is particularly important because expressions of selflessness and related experiences of spiritual transcendence are found in a wide variety of religions and even among nonbelievers. Relatedly, it will be important to gain an understanding of how the perception of our experiences without a sense of self allows one to have a sense of connectedness to all things often described in terms of a unitary consciousness.

We suggest that for future research and methods to become interdisciplinary, developing standardized taxonomies of terms, clearer definitions, and both objective and subjective measures of selflessness will be required. General populations to be studied should include mystics (experts in spiritual transcendence) and individuals from different religious affiliations, spiritual dispositions, and nonbelievers. Neurological studies of persons with disorders of the self, including neurological and psychiatric populations, should also be included to help inform a broader understanding of selflessness.

We argue (as have others) that it is most appropriate for future researchers to conceptualize the self as a neuropsychological process, rather than as a fixed entity. Specifically, we suggest that the "sense of self" can be conceptualized as the locus of perception from which conscious awareness is typically experienced. By conceptualizing how consciousness is experienced through the sense of self, with a progressively diminished sense of self, or without any sense of self, needs to be ascertained. To help understand the manner by which the brain perceives consciousness through the lens of the self, we have utilized empirical research and anecdotal case studies of disorders of the self presented in Chapters 3 and 4. We have also found the ideas of the British philosopher Evan Thompson useful, given his focus on how a sense of self is related to different states of consciousness. In his book *Dreaming, Waking, Being* (2015), Thompson provides a convincing argument that the brain creates a sense of self that changes based on fluctuating causes and conditions. We propose that this sense of self is the result of ongoing neuropsychological processes that create dynamic points of perception from which consciousness is experienced. That is, the brain is (or becomes) hardwired so that experiences are perceived in terms of a self (i.e., I, me, and mine). When this neuropsychological process (i.e., self-orientation) is minimized through injury or behavioral or ritual practices, the sense of self becomes diminished and in extreme instances is lost completely. As a result selfless experiences are perceived, but they are not perceived as relating to a sense of self. In essence, this is often similar to what is described by individuals with severe disorders of the self and often by persons who have had spiritually transcendent experiences (see Chapter 2).

6 THE DISORDERED SELF AND SELFLESSNESS

In order to better understand how a sense of self is constructed, it makes sense to study the experiences of individuals with "disorders of the self." By studying such individuals, it will be possible to increase our understanding of specific neural networks and the neuropsychological processes associated with the manner by which the brain perceives the self. It will be particularly important to determine if there are specific neurophysiological and neuropsychological processes associated with self-orientation. For example, in some disorders of the self, such as in asomatognosia, there is a specific loss in certain aspects of the sense of the physical self. In other self-disorders, such as anosognosia, there is a specific loss of aspects of the psychological self. In delusional misidentification disorders such as Capgras syndrome, there is a specific loss in the sense of the social and emotionally connected self, that is, the relationship between the self and others. The need exists to determine how the physical, psychological, autobiographical, and social senses of the self operate in creating experiences of spiritual transcendence.

When studying individuals with a distorted sense of self, it will be necessary to focus on the subjective nature of their selfless experiences. Most studies of individuals with disorders of the self have focused on describing the neurological *symptoms* associated with such conditions. For example, neurological case studies and research cited in Todd Feinberg's and Julian Keenan's edited volume *The Lost Self* (2005) describe fascinating neurological symptoms associated with different disorders of the self, but little is said about the subjective nature of such experiences. In addition to describing symptoms, it is necessary to better understand the experiences of these individuals who apparently perceive the world with a diminished sense of self. Future research will benefit from using semistructured interviews to clarify the qualitative nature of perceptions of consciousness for those who have a reduced sense of "I," "me," and "mine." How do such individuals describe their physical self? How do they describe their psychological self? How do they describe the experience of self in relationship to other people, the environment, and the divine? By studying these and related neurological disorders, we can learn more about the manner by which consciousness is perceived both with and without a full sense of self, and consequently come to better understand how the extreme loss of the sense of self can lead to experiences of spiritual transcendence.

The nature of the self can also be explored in individuals with psychiatric disorders of the self. For example, Chapter 3 discussed individuals with conditions that involve a distorted sense of the relationship of the self to others, including individuals who believe that family members are impostors (or that impostors are family members), as in Capgras syndrome discussed above. Similarly, individuals with schizophrenia appear to have what has been conceptualized as a shattered or fragmented sense of the self. They have sensory experiences and thoughts but are unable to determine which are coming from themselves directly or emanating from hallucinations. By evaluating such psychiatric disorders of the self, we can better understand how the sense of self processes relationships both to oneself and to others external to the self and possibly gain clues about how we process a sense of self that may include others as reported during certain mystical or near-death experiences.

In addition to studying individuals with neurological disorders to better understand the nature of selflessness and transcendence, future research can create transient neurological "disorders of the self" using modern technologies such as transcranial magnetic stimulation (TMS) (Uddin, Molnar-Szakacs, Zaidel, & Iacoboni, 2006). This technology is also being used to elicit spiritual experiences and potent experiences of spiritual transcendence (Crescentini, Aglioti, Fabbro, & Urgesi, 2014; Crescentini, Di Bucchianico, Fabbro, & Urgesi, 2015). Of particular interest, potential studies can evaluate the specific nature of transcendent experiences induced artificially in individuals from different religious affiliations and nonreligious identities, including those who identify as atheists or as being spiritual but not religious. It will be important to also identify with greater precision which aspects of the self appear to be "turned off" within the brain in order to enhance spiritual connections with the cosmos or the divine and how "selfless" experiences can be produced depending on different contextual factors.

7 FLOW STATES AND THE LOSS OF SELF

It will also be possible to better understand spiritual transcendence by studying other similar experiences that are described in terms of selflessness. For example, research over the past half century has suggested that unique human experiences are associated with a decreased sense of self. For example, Mihaly Csikszentmihalyi and his colleagues (e.g., Csikszentmihalyi, 1990, 2000; Csikszentmihalyi & Csikszentmihalyi, 1992) conducted decades of research on what they called the "flow experience," full of descriptions that show how a person who has become fully immersed in an activity in which they are engaged will often lose their sense of conscious self-awareness. Interestingly, and relevant to our model, research on "flow" experiences has shown that the intrusion of self-consciousness interrupts such states and interferes with experiences of equanimity or relaxation. Flow states produce what Csikszentmihalyi called "optimal experience" where we feel better and learn faster by keeping our self-conscious emotions and our ego-centered sense of self at minimal levels of interference. As Csikszentmihalyi said,

> ... loss of self-consciousness does not involve a loss of self, and certainly not a loss of consciousness, but rather, only a loss of consciousness of self. What slips below the threshold of awareness is the *concept* of self, the information we use to represent to ourselves who we are. *(Csikszentmihalyi, 2008 [1990]; quoted in Ananthaswamy, 2015, pp. 249–250, original emphasis)*

Even children are prone to the types of experiences that Csikszentmihalyi's flow states describe. As the well-known American writer Madeleine L'Engle (1972) has noted,

The concentration of a small child at play is analogous to the concentration of the artist of any discipline. In real play, which is real concentration, the child is not only outside time, he is outside *himself*. He has thrown himself completely into whatever it is he is doing. A child playing a game, building a sand castle, painting a picture, is completely in what he is doing. His *self-consciousness* is gone; his consciousness is wholly focused outside himself. *A Circle of Quiet (1972, Section 1.3, original emphasis)*

More recently, insights about the process of self-focus have been refined and developed through advances in neuroscience based on brain imagining studies. Recent research is helping illuminate the neurophysiological implications of flow states and also how the sense of self (and its attendant loss) operates as neuropsychological processes.

8 THE DMN AND SPIRITUAL TRANSCENDENCE

Although our model of spiritual transcendence has focused on very general neuroanatomical regions (i.e., right hemisphere/parietal lobe), recent research has also implicated other networks that appear to be related to the experience of the self, which need further investigation. Specifically, research has identified an association between the "default mode network" (often referred to as the "DMN") and the ability to focus attention on the self. The DMN is hypothetically engaged when individuals are not engaged in a specific task (i.e., the brain defaults to this network when it is not needed to directly engage with an active mental task in the external world). The DMN is associated with introspection, and studies suggest, for example, that the DMN is activated when individuals are doing nothing and are perhaps daydreaming or their minds are wandering but more typically when ruminating and reflecting in a sort of self-focused and incessant chatter about ourselves. In other words, "the default mode makes each of us the center of the universe as we know it" (Goleman & Davidson, 2017, p. 151). In general, the DMN appears to be related to the default tendency to focus attention on the self.

While still hypothetical as a fully integrated functional system of the brain, the DMN involves activity in a number of neuroanatomical regions in the both hemispheres, including specific areas of the parietal lobe (i.e., especially the inferior parietal lobe and precuneus) that are implicated in both the sense of self in the world and in its detachment from an immediate task. The precuneus of the human brain has been identified as the "functional core of the default mode network" (Utevsky, Smith, & Huettel, 2014), a cerebral network that is highly implicated in human self-introspection and ruminations on the self (see Fig. 1).

The DMN is most active when we are not engaged with (i.e., paying attention to) external stimuli. When this occurs, our mind wanders and

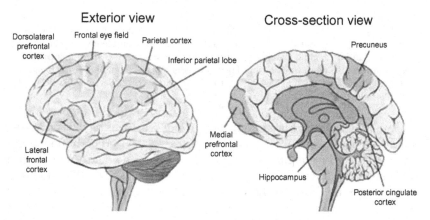

FIG. 1 View of different areas of the human brain including several that are implicated in the DMN.

focuses attention on the self, often ruminating and mulling over our impressions of ourselves when there is little to capture our attention (Goleman & Davidson, 2017), a strong contrast to what occurs in a profound experience of spiritual transcendence when the sense of self is lost.

Research on the DMN seems to indicate that self-related introspection, meaning inward focused attention on one's experience of self, is a separate neurological process from focusing attention on external sensory information. Thus, it appears that it is difficult to place one's attention on the immediate sense of the self and at the same time have awareness of incoming external sensory information. Apparently, the ability to focus attention on one's internal sense of the self (e.g., being lost in thought) differs from one's ability to focus attention on stimuli from the outside world and suggests these are mutually antagonistic processes (Goldberg, Harel, & Malach, 2006) that operate in an inverse relationship with one another (Northoff, Qin, & Feinberg, 2011).

However the inverse relationship between these aspects of the DMN may help explain why the loss of the sense of self experienced during spiritual transcendence often gives the feeling of being connected to everything (i.e., nature, cosmos, universe, and God). Spiritual transcendence may be related to both a reduction in the integration of experiences into a sense of self (i.e., reduced right hemisphere/parietal lobe functioning) and a decreased attention to the ruminating self (i.e., reduced DMN functioning). Research is further suggesting that self-referential processing is more fundamental than perceiving the self as either subject (e.g., "I") or object (e.g., "me") (Northoff et al., 2006). Thus, when these senses of the self are greatly diminished and self is no longer experienced as either subject or object, the groundwork for a potent experience of spiritual transcendence develops. This may explain reports expressed by individuals who have intense experiences of spiritual transcendence that they have both a di-minished sense of self and at the same time a sense of expanded connec-tion with all of creation.

Other research is offering support for the reduced role of the DMN in spiritual transcendence, at least associated with meditation practices. Neuroimaging studies have repeatedly shown, for example, that experi-enced Buddhist meditators show less activity in the DMN than do novices (Brewer et al., 2011; Pagnoni, 2012; Taylor et al., 2012). Interestingly, even when just resting and not actively meditating, highly experienced long-term mindfulness meditators showed less activity and connectivity with the DMN than did comparable nonmeditators (Berkovich-Ohana, Harel, Hahamy, Arieli, & Malach, 2016). Apparently, by regularly engaging in and practicing different forms of meditation (at least in Buddhist types of meditation), connections within the DMN become reduced in general (and not only when one is meditating). These changes from meditation appear to occur as both state (i.e., during meditation) and trait experi-ences (i.e., persisting experiences that occur consistently). Thus, mindful-ness meditation appears to induce changes in neurophysiologic activity in self-referential and attentional networks (Berkovich-Ohana, Glicksohn, & Goldstein, 2011).

Future research may benefit from determining how the ability to focus attention on the self (i.e., self-awareness) associated with the DMN may differ significantly from the right hemisphere's/parietal lobe's ability to integrate different experiences into a sense of self and how together they may lead to spiritual transcendence. Recall that in Chapter 1, we discussed how the precuneus area of the parietal region of the brain has become greatly expanded in recent human evolution. Taken together, these in-sights reinforce our model, for if the evolutionary development of the pa-rietal lobe and the DMN has allowed humans to cultivate an increasingly targeted focus on the self, with the right hemisphere/parietal particularly implicated, our model is suggesting what happens when the opposite oc-curs and the right parietal region becomes increasingly inactivated and focus on the self is reduced greatly or eliminated. A significant loss of the sense of self or selflessness can occur, opening the potential for unique experiences of spiritual transcendence. Borrowing Csikszentmihalyi's ex-pression, perhaps, experiences of spiritual transcendence represent the most powerful expression of flow states.

9 EXPLORING SELFLESSNESS IN DIFFERENT CONTEXTS

In addition to identifying the neurological substrates and neuropsycho-logical processes associated with a diminished sense of self and spiritual transcendence, it is equally important to determine how cultural and religious factors influence such experiences. If the neuropsychological capacity of selflessness is universal for all individuals as we contend, the humanities and social sciences must take the lead in determining how different cultural and religious factors influence the experience and expression of occurrences of selflessness. Future studies may benefit from interviewing experienced practitioners from various religious traditions to find out more about their experiences of a diminished sense of self and the experience of an expanded sense of self where one is connected to something greater than the self. More studies are needed to determine how a wider range of religious and indigenous traditions interpret spiritual transcendence in terms of transitioning from the experience of self-consciousness to the experience of unitary consciousness.

Future research also needs to compare the spiritually transcendent experiences of individuals from different faith traditions to the transcendent experiences of nonbelievers (i.e., atheists, skeptics, and secular humanists). Specifically, are atheists' enhanced experiences of connection with the cosmos similar to experiences of connection with the divine for believers? It will be interesting to determine if and how each of these spiritual experiences may involve a loss of the sense of self.

Finally, research is needed to study the neurophysiology of spiritual experiences of individuals from different faiths (e.g., Hindu, Buddhist, Christian, Muslim, and Indigenous) that are associated with a variety of religious practices (e.g., veneration of a deity, trance, meditation, prayer, chanting, dance, music, and art). Just as Newberg and colleagues demonstrated similar patterns of cerebral blood flow in Buddhists engaged in meditation and Christian nuns engaged in centering prayer (Newberg, Alavi, Baime, Mozley, & d'Aquili, 1997; Newberg, Pourdehnad, Alavi, & d'Aquili, 2003), larger-scale studies may identify neurophysiological similarities and differences in individuals from diverse religious traditions engaging in different religious practices and the spiritual practices of nonbelievers. Such research can further confirm whether there are neurological similarities (i.e., reduced right hemisphere/parietal functioning) in spiritually transcendent experiences regardless of religious or nonreligious orientation, as our research has progressively suggested.

10 OTHER SELFLESS CHARACTERISTICS

One of the most promising areas for future research will be determining the manner by which different aspects of selflessness serve as neuropsychological foundations for a variety of spiritual experiences and other exemplary human characteristics. Hypothetically, it can be argued that empathy, compassion, and altruism may be similarly based in selflessness.

In our previous research, we chose to investigate empathy, given that it has often been described in terms of selflessness. We asked whether it is necessary to reduce focus on oneself in order to relate cognitively and emotionally to the experience of another. Using the same methods as in our previous neuropsychology of spirituality studies, we found the exact opposite of what we expected (Johnstone, Bodling, Cohen, Christ, & Wegrzyn, 2012). The results specifically indicated that *increased* right parietal functioning was associated with *increased* empathy and not the reverse. Based on the manner by which we had been interpreting our results, these findings suggested that one has to have an increasingly *strong* sense of self in order to be empathetic to others. The social psychology of empathy literature provided explanations for our findings, as the social psychologist Cozolino explains:

> In other words, we hold our own perspective in mind while simultaneously imagining what it is like to be the other. In order to have empathy, we need to maintain an awareness of our inner world as we imagine the inner world of the other. *(Cozolino, 2006, p. 203)*

Thus, in contrast to our original hypothesis that empathy is a selfless experience, we found that in order to be empathetic, it is necessary to have a strong sense of self in order to relate to the feelings and thoughts of another as if they were your own. This suggests that empathy is in some ways the opposite of selflessness.

There was an earlier study on the self and empathy that supported these findings, although the authors did not interpret their results in the same manner. Grattan and Eslinger (1989) had also investigated the neuropsychological foundations of empathy in a sample of people with brain dysfunction, and they also found that *increased* right parietal functioning (also measured by the Judgment of Line Orientation test) was associated with *increased* empathy. However, they interpreted their results as suggesting that impairments in visual-spatial perception are potentially related to empathy, which remains hard to explain. However, if the Judgment of Line Orientation test is conceptualized as a measure of the functional integrity of the right parietal lobe, rather than as a measure of spatial perception as they did, their results support our contention that empathy is related to an intact sense of self. Overall, it makes more conceptual sense to relate empathy to an intact sense of self, rather than merely to the ability to perceive spatial relations between line drawings.

In addition to determining relationships between selflessness and transcendence and empathy, similar opportunities exist for future neuropsychological research to identify other spiritually related experiences and human characteristics that may be associated with either an increased or decreased sense of self. These characteristics could include such virtues and characteristics as altruism, compassion, generosity, kindness, gratitude, and humility. This type of research can be strengthened by integrating the insights of religious studies and other areas of the humanities and from the social sciences.

11 SELFLESSNESS, SPIRITUAL TRANSCENDENCE, AND HUMAN FLOURISHING

It is our hope that this book will help stimulate a wide range of academic disciplines to collaborate and to promote selflessness as a way to improve human flourishing, including spiritual well-being. Through a more in-depth understanding of the nature of the self, we believe it will also be possible to enhance religious and spiritual practices and to improve physical, psychological, and social well-being. By using neuroscientific research in conjunction with research from the humanities and social sciences, we may learn how to reduce focus on the self; it may be possible to enhance religious and spiritual practices and ultimately enriched spiritual experiences. Many meditative and prayer practices already promote a reduced focus on the self. Whereas TMS may be used to induce spiritual experiences, biofeedback and neurofeedback can also be used to help individuals learn how to reduce right parietal lobe functioning and the sense of self, ultimately leading to enhanced transcendent experiences.

The application of specially designed selfless interventions also has potential to improve physical and mental health outcomes. Research has consistently shown that spirituality and religiosity are connected to better health outcomes, which has been hypothesized to be related to different factors including divine intervention, healthier habits related to following religious tenets, increased social support from fellow congregants (e.g., lower alcohol consumption and cessation of smoking), more positive outlooks on life (e.g., belief in loving and supportive higher powers), and/or increased placebo effects (Koenig, 2008; Park et al., 2016). Additional research can help to determine if increased selflessness (or a decreased focus on the self) associated with spirituality may be a primary factor leading to improved health in specific contexts.

12 FUTURE APPLICATIONS TO PROMOTE SELFLESSNESS

Undoubtedly, using advanced neuroimaging techniques, the neurosciences will continue to focus on identifying cerebral regions, networks, and neurobiological processes associated with selflessness and spiritual transcendence. Likewise, identifying specific neurophysiological processes associated with different dimensions of the self, such as physical, psychological, and autobiographical selves, will be helpful in this endeavor. Furthermore, it will be important to attempt to identify with greater precision which aspects of the self must be "turned off" or greatly diminished within the brain in order to enhance spiritual connections with the cosmos or the divine. Research using TMS to produce a loss of the sense of self along with technologies that can stimulate or elicit spiritual experiences deserves more attention. Neurofeedback and virtual reality technologies may also be developed to enhance spiritual experiences. Exuberance for and initial production of such technologies is just starting to emerge (Wildman, 2015). The idea, for example, that one could temporarily have, through a virtual reality device, a spiritually transcendent experience may become an eventual reality. The monitoring of muscle tone, skin conductance, heart rate, and brainwaves is already occurring, and just as biofeedback is used to help individuals learn how to control pain and anxiety, specific types of neurofeedback may be developed to help individuals monitor the activity of their brain and learn how to reduce their sense of self in order to have experiences of spiritual transcendence. In addition, neurophysiological research can explore the manner by which specific neurochemicals such as oxytocin, vasopressin, or dopamine may be operating in heightened spiritual experiences.

Although controversial, future research is likely to include increasing exploration of psychoactive substances, what scholars of religion have referred to as "entheogens," namely, different plant substances, like mescaline, peyote, or ayahuasca, that have been used in different cultures. It is noted that plant entheogens have been used for centuries in indigenous religions throughout the world to enhance spiritual experiences. While some scholars of religion have argued that chemically induced experiences are entirely "artificial" (e.g., Zaehner, 1957), others have stated that they can produce genuine divine or cosmic connections (e.g., Huxley, 1954). The well-respected scholar of religions Huston Smith (1964) long ago asserted that a failure to explore the connection between psychoactive substances and spiritual experiences would be akin to theologians' refusal to look through Galileo's telescope because they worried it might change their ideas of humanity's place in the (divine) universe. To ignore these topical territories and the subjective experiences of countless individuals would be to lose access to an important area of research and understanding about the human experience. As William James clearly understood over a century ago,

> ... there lie potential forms of consciousness entirely different. We may go through life without suspecting their existence ... No account of the universe in its totality can be final which leaves these other forms of consciousness quite disregarded. How to regard them is the question—for they are so discontinuous with ordinary consciousness. *(James, 1902, p. 298)*

13 CONCLUDING THOUGHTS

As the title of this concluding chapter indicates, our goal is to build bridges between science, religion, and spirituality. Fitting nicely with our appeal for a more unified approach to selflessness and spiritual transcendence between the sciences and humanities, Thompson makes a strong statement in this direction in his book *Dreaming, Waking, Being* (2015) when he says the following:

> By enriching science with contemplative knowledge and contemplative knowledge with cognitive science, we can work to create a new scientific and spiritual appreciation of human life, one that no longer requires or needs to be contained within either a religious or antireligious framework. *(Thompson, 2015, p. xl)*

If we look to the wisdom of major figures who speak from these different orientations and vantage points, it is apparent that many have already arrived at this important conclusion. As the Dalai Lama, one of the most well-known religious figures of our time, has indicated,

> The spiritual methods are available, but we must make these acceptable to the mass who may not be spiritually inclined. Only if we can do that will these methods have the widest of effect. This is important because science, technology, and material development cannot solve all our problems. We need to combine our material development with the inner development of such human values as compassion, tolerance, forgiveness, contentment and self-discipline. *(His Holiness Dalai Lama, "A Collaboration Between Science and Religion," January 14, 2003)*

By working together and sharing our divergent disciplinary insights with interest and patience for others who may have different perspectives, we can create new types of insights and enhance human understanding through our endeavors. These collaborations can also foster the opportunity to better understand the importance of selflessness in human spiritual experience. As the well-known astrophysicist and cosmologist Carl Sagan understood clearly,

> Science is not only compatible with spirituality; it is a profound source of spirituality. When we recognize our place in an immensity of light-years and in the passage of ages, when we grasp the intricacy, beauty, and subtlety of life, then that soaring feeling, that sense of elation and humility combined, is surely spiritual ... The notion that science and spirituality are somehow mutually exclusive does a disservice to both. *(Sagan, 1996, p. 29)*

References

Alexander, M. P., Stuss, D. T., & Benson, D. F. (1979). Capgras syndrome: A reduplicative phenomenon. *Neurology, 29,* 334–339.

Alvarado, C. S. (2000). Out-of-body experiences. In E. Cardena, S. J. Lynn, & S. Krippner (Eds.), *Varieties of anomalous experiences* (pp. 183–218). Washington, DC: American Psychological Association.

American Psychiatric Association. (2013). *Diagnostic and statistical manual of mental disorders* (5th ed.). Washington, DC: American Psychiatric Association.

Ananthaswamy, A. (2015). *The man who wasn't there: Tales from the edge of the self.* New York, NY: Dutton.

Underhill, E. (Ed.), (1922). *The cloud of unknowing.* Retrieved from http://www.sacred-texts.com/chr/cou/cou49.htm.

Armstrong, K. (2004). *The spiral staircase: My climb out of darkness.* New York, NY: Random House.

Austin, J. H. (1998). *Zen and the brain: Toward an understanding of meditation and consciousness.* Cambridge, MA: MIT Press.

Austin, J. H. (2006). *Zen-brain reflections: Reviewing recent developments in meditation and states of consciousness.* Cambridge, MA: MIT Press.

Austin, J. H. (2009). *Selfless insight: Zen and the meditative transformations of consciousness.* Cambridge, MA: MIT Press.

Barnby, J. M., Bailey, N. W., Chambers, R., & Fitzgerald, P. B. (2015). How similar are the changes in neural activity resulting from mindfulness practice in contrast to spiritual practice? *Consciousness and Cognition, 36,* 219–232.

Barnett, K., Kirk, I., & Corballis, M. (2005). Right hemisphere dysfunction in schizophrenia. *Laterality: Asymmetries of Body, Brain, and Cognition, 10,* 29–35.

Barrett, N. F. (2011). Principles of neurotheology. *Ars Disputandi, 11*(1), 133–136.

Baumeister, R. F., & Exline, J. J. (2002). Mystical self loss: A challenge for psychological theory. *The International Journal for the Psychology of Religion, 12*(1), 15–20.

Baumgardt, D. (1947). Major trends in Jewish mysticism, by Gershom G. Scholem. *Commentary Magazine.* July 1, 1947.

Bear, D. M., & Fedio, P. (1979). Quantitative analysis of interictal behavior in temporal lobe epilepsy. *Archives of Neurology, 34,* 454–467.

Beauregard, M., & O'Leary, D. (2007). *The spiritual brain: A neuroscientist's case for the existence of the soul.* New York, NY: Harper Collins.

Beitman, B. D., & Nair, J. (Eds.), (2004). *Self-awareness deficits in psychiatric patients: Neurobiology, assessment, and treatment.* New York: W.W. Norton.

Benton, A. L., & Hamsher, K. D. (1989). *Multilingual aphasia examination.* Iowa City, IA: AJA Associates.

Benton, A. L., Hamsher, K. D., Varney, N. R., & Spreen, O. (1983). *Judgment of line orientation manual.* New York: Oxford University Press.

Berkovich-Ohana, A., Glicksohn, J., & Goldstein, A. (2011). Temporal cognition changes following mindfulness, but not transcendental meditation practice. *Proceedings of Fechner Day, 27*(1), 245–250.

Berkovich-Ohana, A., Harel, M., Hahamy, A., Arieli, A., & Malach, R. (2016). Data for default network reduced functional connectivity in meditators, negatively correlated with meditation expertise. *Data in Brief, 8,* 910–914.

Bhikkhu, T., (Trans.). (1997). Suflfla sutta: Empty (SN 35.85). In *Access to Insight*. (*Legacy Edition*). Retrieved from http://www.accesstoinsight.org!tipitaka/sn/sn35/sn35.085.than.html.

Bigelow, H. J. (1850). Dr. Harlow's case of recovery from the passage of an iron bar through the head. *American Journal of the Medical Sciences, 20*, 13–22.

Blackmore, S. J. (1982). Out-of-body experiences, lucid dreams, and imagery: Two surveys. *Journal of the American Society for Psychological Research, 76*(4), 301–317.

Blanke, O. (2012). Multisensory brain mechanisms of bodily self-consciousness. *Nature Reviews Neuroscience, 13*(8), 556–571.

Blanke, O., Landis, T., Spinelli, L., & Seeck, M. (2004). Out-of-body experience and autoscopy of neurologic origin. *Brain, 27*, 243–258.

Blanke, O., Mohr, C., Michel, C. M., Pascual-Leone, A., Brugger, P., Seeck, M., et al. (2005). Linking out-of-body experience and self processing to mental own-body imagery at the temporoparietal junction. *Journal of Neuroscience, 25*(3), 550–557.

Blum, J. N. (2014). The science of consciousness and mystical experience: An argument for radical empiricism. *Journal of the American Academy of Religion, 82*(1), 150–173.

Botvinick, M., & Cohen, J. (1998). Rubber hands 'feel' touch that eyes see. *Nature, 391*, 756.

Bouchard, T. J., Lykken, D. T., McGue, M., Segal, N. L., & Tellegen, A. (1990). Sources of human psychological differences: The Minnesota twins reared apart. *Science, 250*, 223–238.

Bouchard, T. J., McGue, M., Lykken, D. T., & Tellegen, A. (1999). Intrinsic and extrinsic religiousness: Genetic and environmental influences and personality correlates. *Twin Research, 2*, 88–98.

Brändström, S., Schlette, P., Pryzbcck, T. R., Lundberg, M., Forsgren, T., Sigvardsson, S., et al. (1998). Swedish normative data on personality using the temperament and character inventory. *Comparative Psychiatry, 39*(3), 122–128.

Breen, N., Caine, D., Coltheart, M., Hendy, J., & Roberts, C. (2000). Towards an understanding of misidentification: Four case studies. *Mind and Language, 15*, 74–110.

Brent, B. K., Seidman, L. J., Thermenos, H. W., Holt, D. J., & Keshavan, M. S. (2014). Self-disturbance as a possible premorbid indicator of schizophrenia risk: A neurodevelopmental perspective. *Schizophrenia Research, 152*, 73–80.

Brewer, J. A., Worhunsky, P. D., Gray, J. R., Tang, Y. Y., Weber, J., & Kober, H. (2011). Meditation experience is associated with differences in default mode network activity and connectivity. *Proceedings of the National Academy of Sciences, 108*(50), 20254–20259.

Bruner, E., Amano, H., Pereira-Pedro, A. S., & Ogihara, N. (2018). The evolution of the parietal lobes in the genus *Homo*. In E. Bruner, N. Ogihara, & H. C. Tanabe (Eds.), *Digital endocasts: From skulls to brains (replacement of Neanderthals by modern humans series)* (pp. 219–237). Tokyo: Springer Japan.

Bruner, E., & Iriki, A. (2016). Extending mind, visuospatial integration, and the evolution of the parietal lobes in the human genus. *Quaternary International, 405*, 98–110.

Bruner, E., & Pearson, O. (2013). Neurocranial evolution in modern humans: The case of Jebel Irhoud 1. *Anthropological Science, 121*(1), 31–41.

Bruner, E., Preuss, T. M., Chen, X., & Rilling, J. K. (2017). Evidence for expansion of the precuneus in human evolution. *Brain Structure and Function, 222*(2), 1053–1060.

Buber, M. (1947). *Between man and man*. R. G. Smith, Trans London: Kegan Paul.

Butler, P. M., McNamara, P., & Durso, R. (2010). Deficits in the automatic activation of religious concepts in patients with Parkinson's disease. *Journal of the International Neuropsychological Society, 16*, 252–261.

Butler, P. M., McNamara, P., & Durso, R. (2011). Side of onset in Parkinson's disease and alterations in religiosity: Novel behavioral phenotypes. *Behavioral Neurology, 24*, 133–141.

Cahalan, S. (2012). *Brain on fire: My month of madness*. New York, NY: Simon & Schuster.

Chiao, J. Y., Harada, T., Komeda, H., Li, Z., Mano, Y., Saito, D., et al. (2009). Dynamic cultural influences on neural representations of the self. *Journal of Cognitive Neuroscience, 22*, 1–11.

Chittick, W.C., (Trans.). (Eds.), (1983). *The Sufi path of love: The spiritual teachings of Rumi.* Albany: State University of New York Press.

Cho, F., & Squier, R. K. (2008). Reductionism: Be afraid, be "very" afraid. *Journal of the American Academy of Religion, 76*(2), 412–417.

Cloninger, C., Pryzbeck, R., Svrakic, T. R., & Wetzel, R. D. (1994). *The temperament and character inventory: A guide to its development and use.* St. Louis, MO: Center for Psychobiology of Personality, Washington University.

Cloninger, C. R., Svrakic, D. M., & Przybeck, T. R. (1993). A psychobiological model of temperament and character. *Archives of General Psychiatry, 50*(12), 975–990.

Feinberg, T. E. (Ed.), (2011, March). Brain and self: Bridging the gap (special edition). *Consciousness and Cognition, 20*(1), 1–172.

Collins, S. (1982). *Selfless persons: Imagery and thought in Theravada Buddhism.* Cambridge, UK: Cambridge University Press.

Costa, P. T., Jr., & McCrae, R. R. (1992). The five-factor model of personality and its relevance to personality disorders. *Journal of Personality Disorders, 6*(4), 343–359.

Cozolino, L. (2006). *The neuroscience of human relationships.* New York, NY: Norton.

Crescentini, C., Aglioti, S. M., Fabbro, F., & Urgesi, C. (2014). Virtual lesions of the inferior parietal cortex induce fast changes of implicit religiousness/spirituality. *Cortex, 54,* 1–15.

Crescentini, C., Di Bucchianico, M. D., Fabbro, F., & Urgesi, C. (2015). Excitatory stimulation of the right inferior parietal cortex lessens implicit religiousness/spirituality. *Neuropsychologia, 70,* 71–79.

Csikszentmihalyi, M. (1990). *Flow: The psychology of optimal experience.* New York: Harper and Row.

Csikszentmihalyi, M. (2000). *Beyond boredom and anxiety.* San Francisco, CA: Jossey-Bass.

Csikszentmihalyi, M., & Csikszentmihalyi, I. S. (Eds.), (1992). *Optimal experience: Psychological studies of flow in consciousness.* New York: Cambridge University Press.

Dalai Lama by His Holiness Tenzin Gyatso, the 14th. (2003, January 14). A collaboration between science and religion. Retrieved from https://www.dalailama.com/messages/buddhism/science-and-religion.

Damasio, A. (2010). *Self comes to mind: Constructing the conscious brain.* New York: Vintage Books.

d'Aquili, E. G., & Newberg, A. B. (1999). *The mystical mind: Probing the biology of religious experience.* Minneapolis, MN: Fortress Press.

Decety, J., & Sommerville, J. A. (2003). Shared representations between self and other: A social cognitive neuroscience view. *Trends in Cognitive Science, 7,* 527–533.

Dein, S., Cook, C. C., Powell, A., & Eagger, S. (2010). Religion, spirituality and mental health. *The Psychiatrist, 34*(2), 63–64.

Delis, D., & Kaplan, E. (2001). *Delis Kaplan executive function system.* San Antonio, TX: The Psychological Corporation.

Dennett, D. C. (1992). The self as a center of narrative gravity. In F. S. Kessel, P. M. Cole, & D. L. Johnson (Eds.), *Self and consciousness: Multiple perspectives* (pp. 103–115). Hillsdale, NJ: Lawrence Erlbaum.

Devinsky, O., & Lai, G. (2008). Spirituality and religion in epilepsy. *Epilepsy & Behavior, 12*(4), 636–643.

Devue, C., & Bredart, S. (2011). The neural correlates of visual self-recognition. *Consciousness and Cognition, 20,* 40–51.

Dewhurst, K., & Beard, A. W. (2003). Sudden religious conversions in temporal lobe epilepsy. *Epilepsy & Behavior, 4*(1), 78–87.

Dittrich, L. (2016). *Patient HM: A story of memory, madness, and family secrets.* New York: Random House.

Ducoff, B. (1989, June 5). *On the ethics of self in Judaism: Searching for the middle ground between selflessness and selfishness.* In: Paper presented at the annual meeting of the Conference of Jewish Communal Service, Boca Raton, FL.

Easwaran, E. (1996/1989). *Original goodness: A commentary on the Beatitudes*. Tomales, CA: Nilgiri Press.

Egan, H. D. (2013). The mystical theology of Karl Rahner. *The Way, 52*(2), 43–62.

Ehrsson, H. H. (2007). The experimental induction of out-of-body experiences. *Science, 317*, 1048.

Einstein, A. (1930). Religion and science. *The New York Times*, (Nov. 9, 1930). Retrieved from http://www.sacred-texts.com/aor/einstein/einsci.htm.

Einstein, A. (1954). The world as I see it. In C. Seelig (Ed.), *Ideas and opinions, based on Mein Weltbild* (pp. 8–11). New York, NY: Bonzana Books [Original work published 1949].

Feinberg, T. E., DeLuca, J., Giacino, J. T., Roane, D. M., & Solms, M. (2005). Right hemisphere pathology and the self: Delusional misidentification and reduplication. In T. E. Feinberg & J. P. Keenan (Eds.), *The lost self: Pathologies of the brain and identity* (pp. 100–130). New York: Oxford University Press.

Feinberg, T. E., & Keenan, J. P. (Eds.), (2005). *The lost self: Pathologies of the brain and identity*. New York: Oxford University Press.

Ferguson, M. A., Nielsen, J. A., King, J. B., Dai, L., Giangrasso, D. M., Holman, R., et al. (2018). Reward, salience, and attentional networks are activated by religious experience in devout Mormons. *Social Neuroscience, 13*(1), 104–116.

Fetzer Institute, National Institute on Aging Working Group. (1999). *Multidimensional measurement of religiousness/spirituality for use in health research*. Kalamazoo, MI: Fetzer Institute.

Forsyth, J. (2003). *Psychological theories of religion*. Upper Saddle River, NJ: Pearson Education.

Frankl, V. E. (1985 [1959]). *Man's search for meaning*. New York: Simon & Schuster.

Frassinetti, F., Maini, M., Romualdi, S., Galante, E., & Avanzi, S. (2008). Is it mine? Hemispheric asymmetries in corporeal self-recognition. *Journal of Cognitive Neuroscience, 20*, 1507–1516.

Freeman, T. (n.d.). What's the difference between Buddhist nothingness and Jewish nothingness? Retrieved from http://www.chabad.org/library/article_cdo/aid/1936781/jewish!Whats-the-Difference-Between-BuddhistNothingness-and-Jewish-Nothingness.htm.

Fuller, R. C. (2001). *Spiritual but not religious: Understanding unchurched America*. New York, NY: Oxford University Press.

Garcia-Romeu, A. (2010). Self-transcendence as a measurable transpersonal construct. *The Journal of Transpersonal Psychology, 42*, 26–47.

Gardner, J. R. (1998). *The developing terminology for self in Vedic India*. [Retrieved from UMI Microform (9834460].

Georgiadis, J. R., Reinders, A. A. T., Paans, A. M., Renken, R., & Kortekaas, R. (2009). Men versus women on sexual brain function: Prominent differences during tactile genital stimulation, but not during orgasm. *Human Brain Mapping, 30*(10), 3089–3101.

Ginsberg, R. (2005). *Kabbalah concepts: Bitul* Retrieved from http://www.mpaths.com/2005/08/kabbalah-concepts-bitul.html.

Giummarra, M. J., Gibson, S. J., Georgiou-Karistianis, N., & Bradshaw, J. L. (2008). Mechanisms underlying embodiment, disembodiment and loss of embodiment. *Neuroscience & Biobehavioral Reviews, 32*(1), 143–160.

Goldberg, I. I., Harel, M., & Malach, R. (2006). When the brain loses its self: Prefrontal inactivation during sensorimotor processing. *Neuron, 50*, 329–339.

Goleman, D., & Davidson, R. J. (2017). *Altered traits: Science reveals how meditation changes your mind, brain, and body*. New York, NY: Penguin Random House LLC.

Goodall, J., & Berman, P. (1999). *Reason for hope: A spiritual journey*. New York, NY: Warner Books.

Grattan, L. M., & Eslinger, P. (1989). Higher cognition and social behavior: Changes in cognitive flexibility and empathy after cerebral lesions. *Neuropsychology, 3*, 175–185.

Greenbaum, A., (Trans.). (2006–2007). Hitbodedut: Meditation and personal prayer. In *The essential Rabbi Nachman*. Retrieved from http://www.azamra.org/essential.shtrnl.

Greyson, B. (2003). Near-death experiences in a psychiatric outpatient clinic population. *Psychiatric Services*, 54(12), 1649–1651.

Greyson, B., Broshek, D. K., Kerr, L. L., & Fountain, N. B. (2015). Mystical experiences associated with seizures. *Religion, Brain, & Behavior*, 5, 182–196.

Griffin, E. (Ed.), (2003). *Modern Spiritual Masters Series*: Evelyn Underhill: Essential writings. Maryknoll, NY: Orbis Books.

Haidt, J. (2012). *Why we love to lose ourselves in religion*. In *TED2012 conference special to CNN updated 10:45 AM EDT, Sun April 1, 2012*.

Hamer, D. (2004). *The God gene: How faith is hardwired into our genes*. New York, NY: Anchor.

Han, S., Gu, X., Mao, L., Ge, J., Wang, G., & Ma, Y. (2010). Neural substrates of self-referential processing in Chinese Buddhists. *Social, Cognitive, and Affective Neuroscience*, 5, 332–339.

Han, S., Mao, L., Gu, X., Zhu, Y., Ge, J., & Ma, Y. (2008). Neural consequences of religious belief on self-referential processing. *Social Neuroscience*, 3(1), 1–15.

Harlow, J. M. (1868). Recovery from the passage of an iron bar through the head. *Publications of the Massachusetts Medical Society*, 2, 327–347.

Harmless, W. (2007). *Mystics*. New York: Oxford University Press.

Harrington, A. & Zajnoc, A. (Eds.), (2003/2006). *The Dalai Lama at MIT*. Cambridge, MA: Harvard University Press.

Harris, S. (2007, January 6). Selfless consciousness without faith. *The Washington Post*. Retrieved from http://www.samharris.org/blog/itern/selflessness-consciousness-without-faith.

Harris, S. (2014). *Waking up: A guide to spirituality without religion*. New York, NY: Simon and Schuster.

Herzog, H., Lele, V. R., Kuwert, T., Langen, K. J., Kops, E. R., & Feinendegen, E. (1990). Changed pattern of regional glucose metabolism during yoga meditative relaxation. *Neuropsychobiology*, 23(4), 182–187.

Holbrook, C., Hahn-Holbrook, J., & Holt-Lunstad, J. (2015). Self-reported spirituality correlates with endogenous oxytocin. *Psychology of Religion and Spirituality*, 7(1), 46.

Holy Bible: The New King James Version. (1982). Nashville, TN: Thomas Nelson.

Hood, R. W., Jr. (2002). The mystical self: Lost and found. *The International Journal for the Psychology of Religion*, 12, 1–14.

Hanson, S. (2018, January 15). *DDD: Living with depersonalization derealization disorder*. Retrieved from https://medium.com/@seanmhanson/ddd-living-with-depersonalization-derealization-disorder-655dcf044cb8.

Hunt, A. N. (2015). Traces of transcendence: C.S. Lewis and the ciphers of being. *Sehnsucht. The C. S. Lewis Journal*, 9, 47–74.

Huxley, A. (1954). *The doors of perception*. New York: Harpers.

James, W. (1902). *The varieties of religious experience*. London, England: Longmans, Green.

James, W. (1984/1892). *Psychology: Briefer course*. Cambridge, MA: Harvard University Press.

John-Julian, O. J. N., Father. (2015). *The complete Cloud of Unknowing: With the letter of privy counsel*. Brewster, MA: Paraclete Press.

Johnstone, B., Bayan, S., Gutierrez, L., Lardizabal, D., Lanigar, S., Yoon, D. P., et al. (2014). Neuropsychological correlates of forgiveness. *Religion, Brain, and Behavior*, 5, 24–35.

Johnstone, B., Bodling, A., Cohen, D., Christ, S. E., & Wegrzyn, A. (2012). Right parietal lobe "selflessness" as the neuropsychological basis of spiritual transcendence. *The International Journal for the Psychology of Religion*, 22, 267–284.

Johnstone, B., Bhushan, B., Hanks, R., Yoon, D. P., & Cohen, D. (2016). Factor structure of the brief multidimensional measure of religiousness/spirituality in US and Indian samples with traumatic brain injury. *Journal of Religion and Health*, 55(2), 572–586.

Johnstone, B., Cohen, D., Christ, S., Wegrzyn, A., & Glass, B. (2014). Functional and structural indices of empathy: Evidence for self-orientation as a neuropsychological foundation of empathy. *Neuropsychology*, 17, 1–11.

Johnstone, B., Cohen, D., Konopacki, K., & Ghan, C. (2016). Selflessness as a foundation of spiritual transcendence: Perspectives from the neurosciences and religious studies. *The International Journal for the Psychology of Religion, 26*(4), 287–303.

Johnstone, B., & Glass, B. A. (2008). Support for a neuropsychological model of spirituality in persons with traumatic brain injury. *Zygon, 43,* 861–874.

Johnstone, B., Hanks, R., Bhushan, B., Cohen, D., Roseberry, J., & Yoon, D. P. (2016c). Selflessness as a universal neuropsychological foundation of spiritual transcendence: Validation with Christian, Hindu, and Muslim traditions. *Mental Health, Religion, and Culture, 20,* 175–187.

Kaplan, J. T., Aziz-Zadeh, L., Uddin, L., & Iacoboni, M. (2008). The self across the senses: The neural response to one's own face and voice. *Social, Cognitive, and Affective Neuroscience, 3,* 218–223.

Kass, J. D., Friedman, R., Lesserman, J., Zutterman, P. C., & Benson, H. (1991). Health outcomes and a new index of spiritual experience. *Journal for the Scientific Study of Religion, 30,* 203–211.

Katz, S. T. (1978). Language, epistemology, and mysticism. In S. T. Katz (Ed.), *Mysticism and philosophical analysis* (pp. 22–74). New York: Oxford University Press.

Kean, S. (2015). *The tale of the dueling neurosurgeons: The history of the human brain as revealed by true stories of trauma, madness, and recovery.* New York: Little, Brown and Company.

Keenan, J. P., Nelson, A., O'Connor, M., & Pascuale-Leone, A. (2001). Self-recognition and the right hemisphere. *Nature, 409,* 305.

Kelsch, C. B., Ironson, G., Szeto, A., Kremer, H., Schneiderman, N., & Mendez, A. J. (2013). The relationship of spirituality, benefit finding, and other psychosocial variables to the hormone oxytocin in HIV/AIDS. *Research in the Social Scientific Study of Religion, 24,* 137–162.

Kilcrease, J. (2014). The bridal-mystical motif in Bernard of Clairvaux and Martin Luther. *The Journal of Ecclesiastical History, 65*(2), 263–279.

Kirk, K. M., Eaves, L. J., & Martin, N. G. (1999). Self-transcendence as a measure of spirituality in a sample of older Australian twins. *Twin Research and Human Genetics, 2*(2), 81–87.

Klein, R. G. (2000). Archeology and the evolution of human behavior. *Evolutionary Anthropology: Issues, News, and Reviews, 9*(1), 17–36.

Klein, R. G. (2017). Language and human evolution. *Journal of Neurolinguistics, 43,* 204–221.

Klosterkotter, J., Hellmich, M., Steinmeyer, E. M., & Schultze-Lutter, F. (2001). Diagnosing schizophrenia in the initial prodromal phase. *Archives of General Psychiatry, 58,* 158–164.

Koenig, H. G. (2008). *Medicine, religion, and health: Where science and spirituality meet.* Philadelphia, PA: Templeton Foundation Press.

Koltko-Rivera, M. E. (2006). Rediscovering the later version of Maslow's hierarchy of needs: Self-transcendence and opportunities for theory, research, and unification. *Review of General Psychology, 10*(4), 302–317.

Lenggenhager, B., Tadi, T., Metzinger, T., & Blanke, O. (2007). Video ergo sum: Manipulating bodily self-consciousness. *Science, 317,* 1096–1099.

L'Engle, M. (1972). *A circle of quiet.* San Francisco, CA: Harper Collins.

Levenson, M. R., Jennings, P. A., Aldwin, C. M., & Shiraishi, R. W. (2005). Self-transcendence: Conceptualization and measurement. *International Journal of Aging and Human Development, 60,* 127–143.

Lewis, G. J., Ritchie, J., & Bates, T. C. (2011). The relationship between intelligence and multiple domains of religious belief: Evidence from a large adult sample. *Intelligence, 39,* 468–472.

Lewis, M., Sullivan, M. W., Stanger, C., & Weiss, M. (1989). Self-development and self-conscious emotions. *Child Development, 60,* 146–156.

Lezak, M. D., Howieson, D. B., & Loring, D. W. (2004). *Neuropsychological assessment* (4th ed.). New York: Oxford Press.

Liu, Y., Sheng, F., Woodcock, K. A., & Han, S. (2013). Oxytocin effects on neural correlates of self-referential processing. *Biological Psychology, 94*(2), 380–387.

Lombardo, M. V., Chakrabarti, B., Bullmore, E. T., Wheelwright, S. J., Sadek, S. A., Suckling, J., et al. (2010). Shared neural circuits for mentalizing about the self and others. *Journal of Cognitive Neuroscience, 22*(7), 1623–1635.

Lou, H. C., Luber, B., Crupain, M., Keenan, J. P., Nowak, M., Kjaer, T. W., et al. (2004). Parietal cortex and representation of the mental self. *PNAS, 101*(17), 6827–6832.

Lutz, A., Dunne, J. D., & Davidson, R. J. (2007). Meditation and the neuroscience of consciousness. In P. D. Zelazo, M. Moscovitch, & E. Thompson (Eds.), *The Cambridge handbook of consciousness* (pp. 499–555). New York, NY: Cambridge University Press.

Lynn, S. J., Berg, J., Lilienfeld, S. O., Merckelbach, H., Giesbrecht, T., Accardi, M., et al. (2012). Dissociative disorders. In M. Hersen & D. C. Beidel (Eds.), *Adult psychopathology and diagnosis* (pp. 497–538). New York: John Wiley & Sons. Chapter 14.

Lynn, R., Harvey, J., & Nyborg, H. (2009). Average intelligence predicts atheism rates across 137 nations. *Intelligence, 37,* 11–15.

Maldonado, J. R., & Spiegel, D. (2008). Dissociative disorders—Dissociative identity disorder (multiple personality disorder). In R. E. Hales, S. C. Yudofsky, & G. O. Gabbard (Eds.), *The American psychiatric publishing textbook of psychiatry.* (5th ed., pp. 681–710). Washington, DC: American Psychiatric Association.

MacDonald, D. A. (2000). *The expressions of spirituality inventory: Test development, validation, and scoring information.* Unpublished test manual (pp. 1–23).

MacMillan, M. (2002). *An odd kind of fame: Stories of Phineas Gage.* Cambridge, MA: The MIT Press.

Maslow, A. H. (1971). *The farther reaches of human nature.* New York, NY: Arkana/Penguin Books.

McNamara, P. (2009). *The neuroscience of religious experience.* New York: Cambridge University Press.

McNamara, P. (2014, August 11). The God effect. *Aeon Newsletter.* Retrieved from https://aeon.co/essays/the-dopamine-switch-between-atheist-believer-and-fanatic.

McNamara, P., Durso, R., Brown, A., & Harris, E. (2006). The chemistry of religiosity: Evidence from patients with Parkinson's disease. In P. McNamara (Ed.), Vol. 2. *Where God and science meet: The neurology of religious experience* (pp. 1–14). Westport, CT: Praeger.

Meador, K. J., Loring, D. W., Feinberg, T. E., Lee, G. P., & Nichols, M. E. (2000). Anosognosia and asomatognosia during intracarotid amobarbital inactivation. *Neurology, 55,* 816–820.

Mercadante, L. A. (2014). *Belief without borders: Inside the minds of the spiritual but not religious.* Oxford/New York: Oxford University Press.

Merton, T. (1949). *Seeds of contemplation.* Norfolk, CT: New Directions.

Merton, T. (1965). *Conjectures of a guilty bystander.* New York: The Abbey of Gethsemani/Random House.

Merton, T. (1983 [1955]). *No man is an island.* New York: Harcourt Brace Jovanovich.

Miller, B.S., (Trans.). (Eds.), (1986). *The Bhagavad Gita: Krishna's counsel in a time of war.* New York, NY: Bantam Books.

Mitchell, R. L. C., & Crow, T. J. (2005). Right hemisphere language functioning and schizophrenia: The forgotten hemisphere? *Brain, 128,* 963–978.

Molnar-Szakacs, I., & Uddin, L. Q. (2013). The emergent self: How distributed neural networks support self-representation. In D. D. Franks & J. H. Turner (Eds.), *Handbook of neurosociology* (pp. 167–182). New York: Springer.

Nakamura, K., Kawashima, R., Sugiura, M., Kato, T., Nakamura, A., Hatano, K., et al. (2001). Neural substrates for recognition of familiar voices: A PET study. *Neuropsychologia, 39,* 1047–1054.

Nelson, B., Whitford, T. J., Lavoie, S., & Sass, L. A. (2014a). What are the neurocognitive correlates of basic self-disturbance in schizophrenia? Integrating phenomenology and neurocognition: Part 1 (source monitoring deficits). *Schizophrenia Research, 152,* 12–19.

Nelson, B., Whitford, T. J., Lavoie, S., & Sass, L. A. (2014b). What are the neurocognitive correlates of basic self-disturbance in schizophrenia? Integrating phenomenology and neurocognition: Part 2 (aberrant salience). *Schizophrenia Research, 152,* 20–27.

Newberg, A. B. (2010). *Principles of neurotheology.* Surrey, England: Ashgate Publishing.

Newberg, A. (2018). *Neurotheology: How science can enlighten us about spirituality.* New York: Columbia University Press.

Newberg, A. B., Alavi, A., Baime, M., Mozley, P. D., & d'Aquili, E. G. (1997). The measurement of cerebral blood flow during the complex cognitive task of mediation using HMPAO-SPECT imaging. *Journal of Nuclear Medicine, 38,* 95.

Newberg, A. B., Alavi, A., Baime, M., Pourdehnad, M., Santanna, J., & d'Aquili, E. G. (2001). The measurement of regional cerebral blood flow during the complex cognitive task of meditation: A preliminary SPECT study. *Psychiatry Research: Neuroimaging, 106,* 113–122.

Newberg, A. B., d'Aquili, E., & Rause, V. (2001). *Why god won't go away: Brain science and the biology of belief.* New York: Ballantine Books.

Newberg, A. B., Pourdehnad, M., Alavi, A., & d'Aquili, E. G. (2003). Cerebral blood flow during meditative prayer: Preliminary findings and methodological issues. *Perceptual and Motor Skills, 97,* 625–630.

Newman, L. E. (1990). Ethics as law, law as religion: Reflections on the problem of law and ethics in Judaism. *Shofar: An Interdisciplinary Journal of Jewish Studies, 9,* 13–31.

Nicholson, R. A., (Trans). (1925–1940). *The Mathnawi of Jalalu'ddin Rumi.* Edited from the Oldest Manuscripts Available: With Critical Notes, Translation & Commentary by Reynold A. Nicholson Litt. D., LL.D., F.B.A. Volume V & VI containing the translation of the third and fourth Books. Retrieved from https://archive.org/stream/RumiTheMathnawiVol5Vol6/Rumi_The-Mathnawi-Vol-5-Vol-6_djvu.txt.

Northoff, G., Heinzel, A., de Greek, M., Bermpohl, F., Dobrowolny, H., & Panksepp, J. (2006). Self-referential processing in the brain: A meta-analysis of imaging studies on the self. *NeuroImage, 31,* 440–457.

Northoff, G., Qin, P., & Feinberg, T. E. (2011). Brain imaging of the self: Conceptual, anatomical and methodological issues. *Consciousness and Cognition, 11,* 52–63.

Olivelle, P., (Trans.). (Eds.), (1998). *The early Upanishads.* New York, NY: Oxford University Press.

Otto, R. (1950/1917). *The idea of the holy* (2nd ed.). New York: Oxford University Press. J. W. Harvey (Trans.).

Pagnoni, G. (2012). Dynamical properties of BOLD activity from the ventral posteromedial cortex associated with meditation and attentional skills. *Journal of Neuroscience, 32*(15), 5242–5249.

Pali Canon (Sayings of the Buddha), *Upadaparitassana Sutta, Samyutta Nikaya* 22.8. Retrieved from http://www.buddha-vacana.org/sutta/samyutta/khandha/sn22-008.html.

Park, C. L., Masters, K. S., Salsman, J. M., Wachholtz, A., Clements, A. D., Salmoirago-Blotcher, E., et al. (2016). Advancing our understanding of religion and spirituality in the context of behavioral medicine. *Journal of Behavioral Medicine, 40*(1), 39–51.

Pew Research Center. (May 12, 2015). *America's changing religious landscape.* Retrieved from http://assets.pewresearch.org/wp-content/uploads/sites/11/2015/05/RLS-08-26-full-report.pdf.

Pew Research Center. (September 6, 2017). *More Americans now say they're spiritual but not religious.* Survey conducted April 25–June 4, 2017. Retrieved from http://www.pewresearch.org/fact-tank/2017/09/06/more-americans-now-say-theyre-spiritual-but-not-religious.

Piedmont, R. L. (1999). Does spirituality represent the sixth factor of personality? Spiritual transcendence and the five-factor model. *Journal of Personality, 67,* 985–1013.

Piedmont, R. L. (2007). Cross-cultural generalizability of the spiritual transcendence scale to the Philippines: Spirituality as a human universal. *Mental Health, Religion & Culture, 10*(2), 89–107.

Piedmont, R. L., & Leach, M. M. (2002). Cross-cultural generalizability of the spiritual transcendence scale in India: Spirituality as a universal aspect of human experience. *American Behavioral Scientist, 45*(12), 1888–1901.

Pike, N. (1992). *Mystic union: An essay in the phenomenology of mysticism.* Ithaca, NY: Cornell University Press.

Poulton, R., Caspi, A., Moffitt, T. E., Cannon, M., Murray, R., & Harrington, H. (2000). Childrens's self-reported psychologic symptoms and adult schizophreniform disorder: A 15-year longitudinal study. *Archives of General Psychiatry, 57,* 1053–1058.

Preilowski, B. (1979). Self-recognition as a test of consciousness in left and right hemisphere of "split-brain" patients. *Activitas Nervosa Superior, 19,* 343–344.

Preyer, W. T. (1889). *The mind of the child: The development of the intellect.* New York: Appleton & Co.

Pusey, E. B. (2013). *The confessions of Saint Augustine, by Saint Augustine Bishop of Hippo.* Project Gutenberg's the confessions of Saint Augustine, by Saint Augustine. Retrieved from https://www.gutenberg.org/files/3296/3296-h/3296-h.htm.

Radovic, F., & Radovic, S. (2002). Feelings of unreality: A conceptual and phenomenological analysis of the language of depersonalization. *Philosophy, Psychiatry, & Psychology, 9,* 271–279.

Ramachandran, V. S., & Rogers-Ramachandran, D. (1996). Synaesthesia in phantom limbs induced with mirrors. *Proceedings of the Royal Society of London, 263*(1369), 377–386.

Reed, P. G. (1991). Self-transcendence and mental health in oldest-old adults. *Nursing Research, 40,* 5–11.

Reitan, R. M. (1992). *Trail Making Test: Manual for administration and scoring.* Tucson, AZ: Reitan Neuropsychological Laboratory.

Roberts, B. (2005). *What is self? A study of the spiritual journey in terms of consciousness.* Boulder, CO: Sentient Publications.

Robinson, J. D., Wagner, N. F., & Northoff, G. (2016). Is the sense of agency in schizophrenia influenced by resting-state variation in self-referential regions of the brain? *Schizophrenia Bulletin, 42,* 270–276.

Rosa, C., Lassonde, M., Pinard, C., Keenan, J. P., & Belin, P. (2008). Investigations of hemispheric specialization of self-voice recognition. *Brain and Cognition, 68,* 204–214.

Ruel, M. (1982). Christians as believers. In J. Davis (Ed.), *Religious organization and religious experience* (pp. 9–31). London, UK: Academic Press.

Ruff, R. L., & Volpe, B. T. (1981). Environmental reduplication associated with right frontal and parietal lobe injury. *Journal of Neurology, Neurosurgery and Psychiatry, 44,* 382–386.

Sacks, O. (1985). *The man who mistook his wife for a hat and other clinical tales.* New York: Touchstone.

Sagan, C. (1996). *The demon-haunted world: Science as a candle in the dark.* New York: Ballantine Books.

Santoni, R. E. (2013). God, creation, and rebellion in Camus: Ambivalent? Inconsistent? Or, finally, incoherent? In N. Jun & S. Wahl (Eds.), *Revolutionary hope: Essays in honor of William L. McBride* (pp. 137–156). Plymouth, UK: Lexington Books.

Sass, L. A. (1998). Schizophrenia, self-consciousness, and the modern mind. *Journal of Consciousness Studies, 5,* 543–565.

Sass, L. A. (2001). Self and world in schizophrenia: Three classic approaches. *Philosophy, Psychiatry, and Psychology, 8,* 252–270.

Sass, L. A. (2014). Self-disturbance and schizophrenia: Structure, specificity, pathogenesis (current issues, new directions). *Schizophrenia Research, 152,* 5–11.

Sass, L. A., & Parnas, J. (2003). Schizophrenia, consciousness, and the self. *Schizophrenia Bulletin, 29,* 427–444.

Saver, J. L., & Rabin, J. (1997). The neural substrates of religious experience. *Journal of Neuropsychiatry and Clinical Neurosciences, 9,* 498–510.

Schjoedt, U. (2009). The religious brain: A general introduction to the experimental neuroscience of religion. *Method and Theory in the Study of Religion, 21,* 310–339.

Scholem, G. (1954). *Major trends in Jewish mysticism* (3rd rev ed.). New York: Schocken Books.

Schreiber, F. R. (1973). *The classic true story of a woman possessed by sixteen separate personalities.* New York: Warner Books.

Seitz, R. J., & Angel, H. F. (2012). Processes of believing—A review and conceptual account. *Reviews in the Neurosciences, 23,* 303–309.

Seymour, J. (2015, May 15). Near to death experience stories of 10 famous celebrities. *The List Café.* Retrieved from http://www.thelistcafe.com/near-death-experience-stories-10-famous-celebrities.

Shrader, D. W. (2008, January). *Seven characteristics of mystical experiences.* In *Sixth annual Hawaii international conference on arts and humanities.*

Siderits, M., Thompson, E., & Zahavi, D. (Eds.), (2011). *Self, no self? Perspectives from analytical, phenomenological, and Indian traditions.* Oxford: Oxford University Press.

Simpson, W. (2014). The mystical stance: The experience of self-loss and Daniel Dennett's "center of narrative gravity". *Zygon, 49,* 458–475.

Slingerland, E. (2008). Who's afraid of reductionism? The study of religion in the age of cognitive science. *Journal of the American Academy of Religion, 76*(2), 375–411.

Smith, H. (1964). Do drugs have religious import? *The Journal of Philosophy, 61*(18), 517–530.

Stace, W. T. (1960a). *Mysticism and philosophy.* London, UK: Macmillan.

Stace, W. T. (Ed.). (1960b). *The teachings of the mystics: Being selections from the great mystics and mystical writings of the world.* New York: Mentor.

Stanton, R. D., Brumback, R. A., & Wilson, H. (1982). Reduplicative paramnesia: A disconnection syndrome of memory. *Cortex, 18,* 23–36.

Sullivan, P. R. (1995). Contentless consciousness and information-processing theories of mind. *Philosophy, Psychiatry, & Psychology, 2,* 51–59.

Sviri, S. (2002). In D. Shlman & G. G. Strouma (Eds.), *The self and its transformation in Sufism: With special reference to early literature* (pp. 195–215). Oxford, UK: Oxford University Press. Self and self-transformation in the history of religions.

Taves, A. (1999). *Fits, trances, & visions: Experiencing religion and explaining experience from Wesley to James.* Princeton, NJ: Princeton University Press.

Taves, A. (2009). *Religious experience reconsidered: A building-block approach to the study of religion and other special things.* Princeton, NJ/Oxford: Princeton University Press.

Taylor, V. A., Daneault, V., Grant, J., Scavone, G., Breton, E., Roffe-Vidal, S., et al. (2012). Impact of meditation training on the default mode network during a restful state. *Social Cognitive and Affective Neuroscience, 8*(1), 4–14.

Thigpen, C. H., & Cleckley, H. M. (1957). *The three faces of Eve.* New York City: McGraw-Hill Education.

Thompson, E. (2015). *Waking, dreaming, being: Self and consciousness in neuroscience, meditation, and philosophy.* New York, NY: Columbia University Press.

Tranel, D., Vianna, E., Manzel, K., Damasio, H., & Grabowski, T. (2009). Neuroanatomical correlates of the Benton facial recognition test and judgment of line orientation test. *Journal of Clinical and Experimental Neuropsychology, 31,* 219–233.

Turner, V. (1969). *The ritual process: Structure and anti-structure.* Chicago: Aldine Publishing.

Uddin, L. Q., Molnar-Szakacs, I., Zaidel, E., & Iacoboni, M. (2006). rTMS to the right inferior parietal lobule disrupts self-other discrimination. *Social Cognitive and Affective Neuroscience, 1,* 65–71.

Underhill, E. (1911). *Mysticism: A study in the nature and development of man's spiritual consciousness.* London, England: Methuen.

Urgesi, C., Aglioti, S. M., Skrap, M., & Fabbro, F. (2010). The spiritual brain: Selective cortical lesions modulate human self-transcendence. *Neuron, 65,* 309–319.

Utevsky, A. V., Smith, D. V., & Huettel, S. A. (2014). Precuneus is a functional core of the default-mode network. *Journal of Neuroscience, 34*(3), 932–940.

Vakoch, D. A. (Ed.), (2012). *Psychology of space exploration: Contemporary research in historical perspective.* Washington, DC: U.S. Government Printing Office.

Van Cappellen, P., Way, B. M., Isgett, S. F., & Fredrickson, B. L. (2016). Effects of oxytocin administration on spirituality and emotional responses to meditation. *Social, Cognitive, and Affective Neuroscience, 11*(10), 1579–1587.

Van Ginneken, V., van Meerveld, A., Wijgerde, T., Verheij, E., de Vries, E., et al. (2017). Hunter-prey correlation between migration routes of African buffaloes and early hominids: Evidence for the "out of Africa" hypothesis. *Integrative Molecular Medicine, 4.* https://doi.org/10.15761/IMM.1000287.

Vásquez, M. A. (2011). *More than belief: A materialist theory of religion.* New York: Oxford University Press.

Swami Vivekananda. (1985–1992). *The complete works of Swami Vivekananda.* Calcutta: Advaita Ashrama.

Wechsler, D. (1997). *WMS-III. Administration and Scoring Manual.* San Antonio, TX: The Psychological Corporation/Harcourt Brace.

Whinnery, J. E., & Whinnery, A. M. (1990). Acceleration-induced loss of consciousness. *Archives of Neurology, 47,* 764–776.

White, F. (1987). *The overview effect: Space exploration and human evolution.* Boston, MA: Houghton Mifflin.

Wiederkehr-Pollack, G. (2007). Self-effacement in the Bible. *Jewish Bible Quarterly, 35,* 179–187.

Wildman, W. (2015 Nov.). *Spirit tech: Understanding technologies of spiritual enhancement, Lecture to the Science and Religion Club.* Columbia, MO: University of Missouri.

Williams, K., & Harvey, D. (2001). Transcendent experience in forest environments. *Journal of Environmental Psychology, 21,* 249–260.

Winquist, C. E. (1998). Person. In M. C. Taylor (Ed.), *Critical terms for religious studies* (pp. 225–238). Chicago, IL: The University of Chicago Press.

Witvliet, C. V. (2001). Forgiveness and health. Review and reflections on a matter of faith, feelings, and physiology. *Journal of Psychology and Theology, (3),* 212–224.

Wu, Y., Wang, C., He, X., Mao, L., & Zhang, L. (2010). Religious beliefs influence neural substrates of self-reflection in Tibetans. *Social, Cognitive, and Affective Neuroscience, 5,* 324–331.

Yaden, D. B., Iwry, J., Slack, K. J., Eiechstaedt, J. C., Zhao, Y., Vaillant, G. E., et al. (2016). The overview effect: Awe and self-transcendent experience in space flight. *Psychology of Consciousness: Theory, Research, and Practice, 3,* 1–11.

Young, A. W., & Leafhead, K. M. (1996). Betwixt life and death: Case studies of the Cotard delusion. In P. W. Marshall & J. C. Marshall (Eds.), *Method in madness: Case studies in cognitive neuropsychiatry* (pp. 147–171). Hove: Psychology Press.

Zaehner, R. C. (1957). *Mysticism, sacred and profane: An inquiry into some varieties of praeternatural experience.* Oxford: Clarendon Press.

Zaehner, R. C. (2016 [1960]). *Hindu and Muslim mysticism.* New York/London: Bloomsbury Publishing.

Zhu, Y., Zhang, L., Fan, J., & Han, S. (2007). Neural basis of cultural influence on self representation. *NeuroImage, 34,* 1310–1317.

Index

Note: Page numbers followed by *f* indicate figures and *t* indicate tables.

Printed in the United States
By Bookmasters